Applied Knowledge Test
for the new **MRCGP**

Questions and Answers for the AKT

Nuzhet A-Ali MBBS, MRCGP, DRCOG, DFFP, DCH

GP in Berkshire

NEWMMRCGP

Scion

First published 2008

A CIP catalogue record for this book is available from the British Library.

ISBN 978 1 904842 54 5

Scion Publishing Limited
Bloxham Mill, Barford Road, Bloxham, Oxfordshire OX15 4FF
www.scionpublishing.com

Important Note from the Publisher
The information contained within this book was obtained by Scion Publishing Limited from sources believed by us to be reliable. However, while every effort has been made to ensure its accuracy, no responsibility for loss or injury whatsoever occasioned to any person acting or refraining from action as a result of information contained herein can be accepted by the authors or publishers.

The reader should remember that medicine is a constantly evolving science and while the authors and publishers have ensured that all dosages, applications and practices are based on current indications, there may be specific practices which differ between communities. You should always follow the guidelines laid down by the manufacturers of specific products and the relevant authorities in the country in which you are practising.

Although every effort has been made to ensure that all owners of copyright material have been acknowledged in this publication, we would be glad to acknowledge in subsequent reprints or editions any omissions brought to our attention.

Typeset by Phoenix Photosetting, Chatham, Kent, UK
Printed by Gutenberg Press Ltd, Malta

Applied Knowledge Test
for the new MRCGP

Contents

Preface

The membership of the Royal College of General Practitioners has recently undergone great changes and I have therefore written this book to assist and encourage all those sitting the nMRCGP Applied Knowledge Test.

The book aims to cover as much of the examination syllabus as possible and uses the same forms of questioning as the actual examination. It contains over 300 questions of single best answer, extended matching, picture and algorithm style. I have given explanations to aid understanding and also provided up-to-date references from the *BMJ*, *BJGP*, *Drugs and Therapeutics Bulletin, NICE/SIGN Guidelines*, and the *BNF*.

Taking into consideration the feedback given by examiners regarding the candidates taking the first AKT in October 2007, I have included those areas which were noted to have caused particular difficulty for candidates, such as care of the acutely ill patient / loss of consciousness, sexual health, sickness certification, and ENT. Examiners' comments on previous MCQ papers of the MRCGP raised concerns about lack of knowledge of anaphylaxis management, meningococcal disease, dermatology, glycaemic index, barrier contraception, influenza, and neonatology, and so these areas have also been covered.

I would like to thank my teachers past and present for their encouragement, and my family and friends for their support.

This book is dedicated to Amina, Saira and Miriam.

Good luck!

Dr Nuzhet A-Ali
January 2008

Acknowledgements

I would like to thank the reviewers of early drafts of this book: Dr Julia Fisher, Dr Penny Halls, Dr Craig Mason, and Dr Kate Roberts-Lewis, for their helpful comments and critiques.

An introduction to the applied knowledge text (AKT)

The nMRCGP

The main difference between the nMRCGP and the MRCGP is that the nMRCGP is now an exit exam without which you cannot practice as a GP. From August 2008 the endpoint assessment for GP training will be the nMRCGP; satisfactory completion of the training and assessment scheme is necessary to obtain the Certificate of Completion of Training in General Practice and will be a requirement for entry on to the General Medical Council's GP Register.

This book is intended to help candidates identify learning needs, prioritise reading and practise questions with a view to passing the Applied Knowledge Test part of the new MRCGP.

The second component of the exam is the Clinical Skills Assessment (similar to the old Simulated Surgery); this is an assessment of a doctor's ability to integrate and apply clinical, professional, communication and practice skills in general practice; it is organized nationally in an assessment centre (currently based in Croydon) and includes 13 simulated patient consultations using actors and objective structured clinical examination-type stations. The book, *Cases and Concepts for the new MRCGP* by Prashini Naidoo, will assist your preparation for this section of the exam.

The third component of the exam is the Workplace Based Assessment. This consists of the Enhanced Trainers Report and also a number of externally moderated assessment tools such as 360 degree feedback.

The AKT

The AKT is designed to test the application of knowledge and the interpretation of information. This was first introduced in October 2007 and replaces the multiple choice paper of the MRCGP and that of summative assessment.

Candidates may sit the AKT at any time during their training, but the RCGP has advised that candidates would be best suited to take it while working as a GPStR during their third year of GP speciality training when they will actually be working in General Practice (ST3).

The AKT takes the form of a a three hour exam comprising 200 single best answer, extended matching, algorithm completion and picture questions. The question breakdown is approximately 80% clinical, 10% administration and 10% evidence-based medicine and critical appraisal; other question formats include table completion and fill-in-the-gaps.

The exam can currently be taken at any one of three sittings a year in October, January and May. Candidates apply to sit the test online via the RCGP website (www.rcgp.org.uk) and then telephone Pearson VUE (the telephone number to call is provided in the automatic email that confirms acceptance of your application) to choose a test centre. Obviously, the earlier a candidate books, the more likely they will be able to sit the exam at their preferred centre.

The exam is computer-based but Pearson VUE will provide a pen and wipe-clean white board to each candidate for workings-out; candidates will not be allowed to take pen and paper into the exam with them.

Do visit the Pearson VUE website (www.pearsonvue.com/rcgp) and make yourself familiar with the layout of a typical centre as well as take a tutorial in computer-based testing.

Having been used to taking previous exams using pen and paper, the computer-based tests may appear disconcerting and it is worth discussing the experiences of colleagues who have already done the exam. Someone I spoke to recently said that the strangest aspects were the cameras and microphones in the room, including one positioned above his head!

One good thing, however, is that because there are 150 Pearson VUE centres throughout the UK and Northern Ireland, there is bound to be one conveniently located near you.

There are no limits to the number of times the AKT may be attempted (though it is £360 per sitting) and, as before, there is no negative marking so, if in doubt, make an educated guess!

Exam preparation

My main advice here would be to practise as many questions as you can – for those questions you get wrong, see this as an educational gap and try to fill it – and read around the subject using journals such as the *BJGP* and *BMJ*; other good sources of information are the *Drugs and Therapeutic Bulletin* and the *BNF*. Also, make use of the internet and visit sites such as NICE and SIGN; Bandolier is also useful. The sacred texts by Neighbour, Pendleton, Berne and Balint are not only helpful for the exam, they are also quite interesting.

You should use the questions in this book to guide your studying rather than trying to read everything ever written and then attempt the questions. Beware of concentrating your reading on subjects you already know well. As Confucius said:

> *If you know, recognize that you know,*
> *If you don't know, realize that you don't know,*
> *That is Knowledge.*

I would also suggest looking at the RCGP website to familiarize yourself with the GP curriculum on which the exam is based and also to see examples of questions; the examiner's comments on previous MCQs are still valid for this exam and are well worth reading.

Certain topics such as statistics, epidemiology, sick certification, mental health Sections, benefits and DVLA exclusions lend themselves to MCQ and so these should be known well. Candidates have often said they find practice management issues difficult to revise: I would suggest spending some time with all the members in the primary care team to understand what their roles are, but especially the practice manager: you can ask them to explain concepts such as risk management, employment law, etc. Make the most of your year in general practice and get involved in practice issues by attending practice meetings.

It is helpful to get together with a group of colleagues to go over question papers together, share ideas and support one another. You may well remember something you talked about better than something you read.

The exam is written by practising general practitioners and based on everyday practice – it helps to keep a diary of PUNS and DENS so that you can look up areas of weakness and make notes as you go along; this is also a good habit to get into for appraisal later on.

The night before the exam, relax with a warm bath; go to bed early and get a good night's sleep; for those of you with responsibilities such as small children, ask someone else to look after them for the night. Make sure you know where you are going and what time you need to be there: ideally do a practice run of your route beforehand and have a look at the building where you'll be tested: familiarity with the venue is always helpful. On the day of the exam avoid too many stimulants: you will be wired enough as it is and extra caffeine will only make you agitated and cause a diuresis.

Passing the exam is not just about knowledge base, but also exam technique. Familiarize yourselves with terms such as pathognomonic, diagnostic, frequently, significantly, characteristically. Make sure you read the instructions for each question carefully: look out for negatives (e.g. which one of the following is NOT a side effect?) before clicking the mouse on the correct box; if you don't know the answer to a question or are getting bogged down, make an educated guess, as gut feeling is often right. Most importantly, keep an eye on the time – a timer on the screen will tell you how much you have left. At the end, check through your work and then walk away; post mortems are never helpful.

Best of luck!

Questions 1-50

for answers see pages 13–20

1. Dementia and driving

A 72 year old man presents with a five month history strongly suggestive of Alzheimer's disease having scored 22/30 on MMSE; this is confirmed by the local psychogeriatrician. You know that he is driving.

Which one of the following is true?
A   GP must inform the DVLA immediately
B   There is a very high risk of car crashes in drivers with dementia in the first few years of presentation
C   Licences are normally valid up to the age of 70 years
D   GP must decide if the patient is fit to drive
E   DVLA states patients with mild, early dementia must not drive

2. Faecal incontinence

Which one of the following is true?
A   Prevalence of faecal incontinence in the community is 15%
B   Rectal bleeding, unexplained changes in bowel habit and anaemia should be investigated routinely
C   Obstetric history may be relevant
D   Loperamide hydrochloride is the second line treatment of choice for faecal incontinence associated with loose stools, when appropriate investigations and treatment have failed to resolve loose stool

3. BNF symbols

Which one of the following is true?
A   The black inverted triangle identifies drugs that have been licensed within the past year
B   The black inverted triangle signifies drugs that are exempt from MHRA's Yellow Card scheme
C   The black inverted triangle denotes newly licensed drugs that are being intensively monitored by MHRA
D   The black triangle denotes drugs that are soon to be off patent and are therefore likely to become cheaper
E   The black triangle denotes those drugs that can be bought over the counter

4-8. Back pain

A   Mechanical low back pain
B   Symphysis pubis dysfunction
C   Osteoporotic vertebral fracture
D   Ankylosing spondylitis
E   Spinal claudication
F   Neoplasm
G   Lumbar disc prolapse with sciatic nerve entrapment

For each scenario depicted below, select the single most likely diagnosis from the list above. Each stem may be used once, more than once, or not at all.

4.   A 23 year old trainee chef presents with a 4 month history of back pain and stiffness; this is always worse in the mornings and as a result he is often late for work; investigations show a raised ESR and you recall seeing his father in the past for back problems and uveitis.

5.   A 33 year old primigravida in her 20th week of pregnancy presents with sacroiliac and hip pain; she reports difficulty getting in and out of her car and a grinding sensation in the pubic area.

6.   A 62 year old lady presents with a constant gnawing pain localized to her thoracic spine; she was treated 2 years ago for breast cancer and is found to have a raised ESR. She notes her clothes are becoming looser.

7.   You are called to the home of a slim 58 year old lady who is complaining of acute, severe back pain; she is very tender over the thoraco–lumbar junction but there is no bruising visible; pain is worse if she coughs, when it radiates around ribs and waist to front; she recalls tripping slightly on the rug as she reached for her cigarettes. You note she was a ballerina in her youth.

8.   A 38 year old primary school teacher was helping her husband lay a patio over the weekend; she awoke the following morning with acute pain over her lower back, made worse by moving; she took some rest and the pain has responded well to paracetamol and ibuprofen; she is wondering whether to go back to work.

9. Carpal tunnel syndrome

Which one of the following is not associated with carpal tunnel syndrome?
A   Myxoedema
B   Diabetes mellitus
C   Acromegaly
D   Rheumatoid arthritis
E   Diabetes insipidus
F   Pregnancy
G   Amyloidosis

10. Cystic fibrosis

Which one of the following is true?
A   1 in 50 adults in the UK carry the CF gene
B   CF is equally common in all races
C   Positive sweat test is diagnostic
D   Patients require additional vitamins B, C and E
E   Boys and girls grow up to be infertile
F   10% of adults develop glucose intolerance

11-14. Heavy menstrual bleeding

Where no structural or histological abnormality is suspected, or revealed on examination and investigation, *which of the following (one in each case) would be most appropriate for the patients described below?*

A   Progesterone intrauterine system
B   Combined oral contraceptive pill
C   Low dose oral progesterones during luteal phase of menstrual cycle
D   Non-steroidal anti-inflammatory drugs
E   Tranexamic acid
F   Hysterectomy
G   Dilation and curettage

11. 27 year old housewife had uncomplicated delivery 8/12 ago but experiencing regular heavy periods with flooding and clots since six months; not planning more children for at least another 2 years.

12. 18 year old student in steady relationship finding heavy bleeding difficult to cope with and interfering with studies. Smokes cigarettes, 5–10/day.

13. Heavy painful periods in 25 year old secretary; not in a relationship at moment, no contraceptive requirements. Fhx thrombophilia.

14. 42 year old lady with heavy periods and needing to use double sanitary protection; doesn't want hormones but does want something effective to make her bleeding lighter.

15. Heavy menstrual bleeding

Which one of the following investigations is recommended when a woman first presents with the problem of heavy bleeding during menstruation?

A Full blood count
B Serum ferritin
C Thyroid function tests
D Hormone profile
E Quantitative assessment of menstrual blood flow
F Saline infusion sonography

16. The menopause

All of the following parameters rise following the menopause, except which one?

A Ferritin
B FSH
C ESR
D Billirubin

17. Temporary residents

Temporary residents are correctly defined as living in the practice area for which one of the following periods?

A Less than 24 hours
B More than 24 hours, less than 1 week
C More than 72 hours, less than 1 week
D More than 24 hours, less than 72 hours
E More than 24 hours, less than 3 months

18. Practice leaflet

The practice leaflet must do all of these things except which one:

A Be updated annually
B Indicate access for disabled people
C Indicate arrangements for dispensing drugs and repeat medication
D State names and numbers of local pharmacies
E State name and address of local walk-in centre

19. The Caldicott Guardian

The Caldicott Guardian's role is best described in terms of which one of the following:

A   Patient confidentiality
B   Communicable disease control
C   Personnel security
D   Patient complaints
E   Practice accounts

20. Colic

With regard to colic, which one of the following statements is true:

A   Affects 60% of babies
B   Tends to occur most commonly in first born males
C   Research shows that babies with colic tend not to eat or gain as much weight as their peers
D   Symptoms always occur during the evening
E   Rare before 2–4 weeks and can last 3 months or more

21. Systemic lupus erythematosus

The following factors are known to make SLE active, except which one?

A   Sunshine
B   Pregnancy
C   Stress
D   Infection
E   Steroids
F   Isoniazid

22-25: Paediatric orthopaedics

A   Transient synovitis of the hip
B   Slipped upper femoral epiphysis
C   Septic arthritis
D   Perthe's disease

Match the following presentations with the single most appropriate diagnosis from the list above.

22. 2 year old boy was walking well from 18 months of age, but over past two weeks is refusing to weight-bear and is complaining of 'leg hurting' and wants to be carried all the time.

23. 7 year old boy attends with mum giving one month history of pain in both knees; limping.

24. 14 year old boy presenting with pain at rest in groin, both sides; shy, overweight; limited abduction and medial rotation on examination; leg appears shortened and externally rotated.

25. 4 year old boy presenting with pain in right hip; lying quietly and very still; reluctant to be examined, systemically unwell.

26-32: Ophthalmology

A Amourosis fugax
B Hyperthyroidism
C Retinal vein occlusion
D Uveitis
E Herpes zoster
F Herpes simplex
G Optic neuritis
H Vitreous haemorrhage
I Foreign body

Match the following presentations with the single most appropriate diagnosis from the list above. Each stem may be used once, more than once, or not at all.

26. A 56 year old businessman who smokes heavily attends worried because he noticed a brief loss of vision, like a curtain coming over his right eye, whilst driving this morning; sight has returned to normal by the time he sees you.

27. A 25 year old mature student who is being seen by one of your partners for chronic low back pain and morning stiffness attends with acute onset of pain affecting the left eye, photophobia, blurred vision and watering of the eye.

28. A 67 year old gentleman presents with a blistering rash affecting his right upper eye lid and part of his forehead; the rash was preceded by a painful tingling sensation.

29. A 56 year old gentleman, who is known to have mixed hyperlipidaemia, presents having experienced sudden loss of vision affecting the right eye; on fundoscopy, the fundus looks like a stormy sunset with haemorrhages and engorged veins.

30. A 32 year old baker presents with pain, photophobia, blurred vision and watering of the eyes; when stained with fluorescein and viewed through a blue light, you notice what looks like a linear branching pattern on the surface of the eye. There is no history of trauma.

31. A 60 year old lady with a 20 year history of diabetes attends with a sudden, painless loss of vision; despite repeated adjustments to the lenses of your ophthalmoscope, you are unable to visualize the retina.

32. A 24 year old lady attends complaining of a gradual loss of colour discrimination over the past few weeks; eye movements are uncomfortable; a few weeks previously she experienced problems with her speech; six months ago she had an episode of pins and needles affecting her forearm which settled spontaneously by the time she saw the doctor.; she is a teacher in primary school and under a lot of stress at the moment.

33. Neurology

A 40 year old lorry driver attends complaining of numbness and weakness affecting his lower limbs; this started a few days ago in his lower legs but is progressing upwards and he has difficulty coming upstairs to your surgery room; he smokes 5–10 cigarettes per day and drinks 3 units of alcohol per week. He was last seen at the surgery 3 weeks ago with a viral sore throat for which he was advised to take paracetamol. On examination, he has absent ankle and weak knee reflexes.

Which one of the following is the most likley diagnosis?
A Acute idiopathic polyneuritis
B Cauda equina syndrome
C Vitamin B6 deficiency
D Vitamin B12 deficiency
E Hansen's disease

34. Neurology

You are called to the home of an elderly lady who has been found collapsed next to her bed by her carer; she opens her eyes only when you ask her to, and offers her arm when you ask to do her blood pressure; her speech consists of inappropriate words.

Her Glasgow coma scale score is:
A Twelve
B Thirteen
C Fourteen
D Fifteen
E Sixteen

35. Erectile dysfunction

Treatment on the NHS is available for all patients with erectile dysfunction except which one of the following groups?
A Those with diabetes mellitus
B Those with multiple sclerosis
C Those with Parkinson's disease
D Those already receiving treatment prior to 14/8/1999
E Those with a single gene neurological disease

36–39. Physical examination

A 34.7 – 37.3
B 35.5 – 37.5
C 35.8 – 38.0
D 36.6 – 38.0

For each of the places of measurement below, select the single most appropriate normal temperature range in degrees celsius from the list above.

36. Ear

37. Mouth

38. Axilla

39. Rectum

40. Venlaflaxine

Which one of the following statements regarding venlaflaxine is correct?
A Is considered superior to cognitive behavioural therapy in the management of mild depression
B Is the treatment of choice in the management of depression in patients with recent myocardial infarction or unstable angina
C Requires baseline ECG in all patients before initiating treatment and during
D Is an SNRI

41-46: Urology

A Diabetes mellitus
B Renal calculus
C Nephrotic syndrome
D Acute pyelonephritis
E Transitional cell carcinoma of bladder
F Chronic renal failure
G Munchausen syndrome
H Diabetes insipidus
I Cystitis

Match the following presentations with the single most appropriate diagnosis from the list above. Each stem may be used once, more than once, or not at all.

41. 15 year old boy attends with mum complaining of fever and rigors; examination reveals mild left loin tenderness; dips urine shows turbid urine, positive for nitrites and leukocyte esterase.

42. 20 year old girl attends with increased appetite, thirst over past few weeks; dip urine shows positive for ketones and glucose; specific gravity is 1.010. She thinks that her clothes feel looser.

43. 4 year old boy attends with mum who is worried that he seems very lethargic over past couple of weeks and now she has also noticed he is looking puffy around the eyes; urine is cloudy, frothy, negative to nitrites and blood but positive to protein+++.

44. 50 year old man requests home visit after a sleepless night with intermittent waves of excruciating lower abdominal pain; on examination he is writhing around the bed, finding it difficult to get comfortable. Urine is cloudy with blood+++; negative to leukocyte esterase, nitrites, protein; microscopic urinalysis shows wbc 2–5/hpf, rbc >100/hpf.

45. A 21 year old catering assistant attends having slipped on a wet floor at work and landed awkwardly on his side; he has brought a urine sample with him. The sample is clear, red and negative to blood, glucose, protein, leukocyte esterase.

46. 60 year old man attends complaining of mild dysuria and occasional hesitancy for past few weeks; he is otherwise well, takes no medication, drinks 14 units of alcohol per week and smokes 40 cigarettes/day. Urine is positive to blood++ and there is a trace of protein, trace of ketones; urine is sent for microscopy and results return showing wbc<2/hpf, rbc 10–30/hpf, occasional hyaline casts and presence of atypical uroepithelial cells.

47-50: Target INR

A   2.0–3.0
B   2.5–3.5
C   3.0–4.0

Which one of the INR ranges given above is appropriate for each of the following diagnoses?

47. Antiphospholipid syndrome

48. First below knee vein thrombosis and no persistent risk factors

49. First proximal vein thrombosis and no persistent risk factors

50. First generation mechanical prosthetic heart valve

Answers to questions 1-50

1. Answer C is true.

There is a growing population of elderly drivers in this country and a clinical review in June 2007 highlighted this (Breen *et al. BMJ*, 2007; **334**: 1365–9). Many people with early dementia can drive safely and the risk of crashes remains low up to three years in a large proportion; however, it is important to reach a balance between independence and safety.

The DVLA (www.dvla.gov.org) *At A Glance Guide* (revised in Sept 2007) states that "group 1 licences are normally valid up to age 70; there is no upper age limit but after age 70 renewal is necessary every 3 years; all licence applications require a medical self declaration by the patient." It is then the legal responsibility of the DVLA to decide whether someone is medically unfit to drive.

It is the responsibility of the licence holder to disclose any medical condition that may impair his/her ability to drive; the GP must therefore make the patient aware of this at the time of diagnosis; he should repeat the advice verbally and in writing as necessary, and in the event that the patient continues to drive despite being warned of the dangers, or the patient does not understand (e.g. dementia), the GP may choose to breach confidentiality and inform the DVLA directly, having advised the patient that this is what he intends to do.

In early dementia, when sufficient skills are retained and progression is slow, a licence may be issued subject to annual review; a formal driving assessment may be necessary. However, those with poor cognition, disorientation and lack of insight or judgement are almost certainly not fit to drive (page 30 in DVLA *At a Glance Guide*).

2. Answer C is true.

NICE (*2007*) suggest that prevalence in the community is nearer 1–10%.

Symptoms suggestive of a lower GI cancer that warrant urgent referral include:
- palpable intraluminal rectal mass on DRE
- rectal bleeding and change in bowel habit for >6/52 in someone >40 years
- rectal bleeding and/or change in bowel habit for 6 weeks in someone >60 years
- RLQ abdominal mass suggestive of large bowel involvement
- unexplained Fe deficiency anaemia

(*NICE Cancer Referral Guidelines, 2005*).

Women who have recently given birth are at high risk of faecal incontinence, especially if they have had a significant obstetric injury.

Loperamide is an anti-motility drug that works on opioid receptors in the bowel, reducing peristalsis, and is the drug of first choice.

3. Answer C is true.

The black triangle symbol denotes drugs that are newly licensed and are being intensively monitored for reported adverse reactions by the Medicines and Healthcare Products Regulatory Agency; yellow cards may be used to report such reactions and can be found at the back of the *BNF*, but reactions may also be reported online and a new pilot scheme has been introduced to encourage self-reporting by patients, parents and carers.

There is no standard time for which products retain a black triangle; safety data are usually reviewed after two years.

4. Answer D is correct.

Ankylosing spondylitis is associated with HLA B27; early X-rays show widening of SI joints and marginal sclerosis; later, fusion of SI joints and vertebral squaring and fusion (a.k.a. bamboo spine).

5. Answer B is correct.

Symphysis pubis dysfunction – due to the effect of pregnancy hormones on pelvic joints leading to inequalities in movement – sometimes causes quite disabling pain anywhere around the pelvic girdle; treatment is supportive with rest, painkillers, physiotherapy; usually resolves after delivery, may recur with subsequent births.

6. Answer F is correct.

Presentation of back pain in anyone under 20, or over 55, that has a constant, progressive, non-mechanical nature, and where there is a past history of cancer, is one of the so-called red flags for potentially serious spinal pathology and warrants urgent (<4/52) referral.

Other red flags are:
- history of HIV, systemic steroid, drug abuse
- systemically unwell, weight loss
- violent trauma
- thoracic pain, widespread neurology

7. Answer C is correct.

Other risk factors for osteoporotic fractures include systemic steroid use, family history of osteoporosis; encourage patients at risk to eat well and maintain BMI >19 kg m^{-2}, to take regular weight-bearing exercise, stop smoking, and take alcohol only within set limits; calcium and vitamin D supplements may be beneficial.

8. Answer A is correct.

The average GP sees >50 acute backs per year; 15 million work days are lost each year due to mechanical low back pain; most resolve within a few weeks; simple analgesia and early mobilization are recommended in most patients.

9. Answer E is correct.

An easy way to remember the causes of carpal tunnel syndrome is: MEDIAN TRAP

Myxoedema
o**E**dema
Diabetes mellitus
Idiopathic
Acromegaly
Neoplastic

Trauma
Rheumatoid arthritis
Amyloidosis
Pregnancy

10. Answer C is correct.

1 in 25 adults in the UK carry the CF gene which is inherited in an autosomal recessive manner; it is most common in Caucasians, and rare in people of Afro–Caribbean origin. If both parents are carriers, there is a 1 in 4 chance of their offspring being affected.

There is a single gene mutation resulting in abnormal function of the CF conductance regulator; this is essential for trans cell membrane salt and water movement, resulting in thickened, salty secretions in the lung, gut and reproductive organs. Consequently, boys are azoospermic and infertile; girls, however, grow up to be sub-fertile and conception is possible.

Patients require pre-meal oral pancreatic enzymes (Creon), a high calorie diet, and supplements of fat-soluble vitamins A, D, and E.

The risk of impaired glucose tolerance is significant; the cumulative incidence of diabetes mellitus in a study in Denmark was 24% in patients aged 20, increasing to 76% in those age 30 (Lanng *et al. BMJ*, 1995; **311**: 655–9).

11. Answer A is correct.

LNG-IUS (Mirena) is suggested as the first-line treatment to be offered provided long-term use (>12/12) is anticipated; it is very effective and decreases blood loss by 90% when used over 12 months.

12. Answer B is correct.

Oral progestogens during the luteal phase are used by some GPs but are not recommended as they are not very effective when given at the usual low dose; they are considered third-line after COCP and non-hormonal methods (they can be used at higher doses of 5 mg t.d.s. starting 3 days before expected date of menstruation to postpone bleeding, but that is a different clinical scenario).

13. Answer D is correct.

NICE Clinical Guideline on Heavy Menstrual Bleeding (*CG44*, January 2007) states that when HMB coexists with dysmenorrhoea, NSAIDs should be preferred to tranexamic acid.

14. Answer E is correct.

Tranexamic acid is an anti-fibrinolytic agent and can reduce blood flow by 50%. Mefanamic acid (Ponstan) reduces bleeding by only 25–30% but is better for those with dysmennorhoea. The other options are hormonal in nature; surgery would not be offered first-line.

> 📋 Note that D&C is an investigation, not a treatment!

15. Answer is A.

NICE Guidance (2007) does not recommend investigations other than FBC.

16. Answer is D.

ESR rises with increasing age.

17. Answer is E.

Anyone in the practice area for <24 hours, treated as an emergency or immediately requiring treatment; anyone residing for more than 24 hours and less than 3 months, as a temporary resident.

18. Answer is D

The practice leaflet is not obliged to state the details of local services such as pharmacies, dentists, etc. but many practices do so.

19. Answer is A.

The Caldicott Report was issued in December 1997 to help deal with concerns about the way patient information is used in the NHS, and due to concerns about confidentiality.

A senior health person should be nominated as 'the Guardian' in each health organization and it is their remit to be responsible for safeguarding the confidentiality of patient-identifiable information, e.g. by using NHS number rather than name of patient wherever possible.

20. Answer is E.

Colic is diagnosed when there is uncontrolled, extended crying in an otherwise healthy baby for longer than three hours every day for more than three days in a week; it is the extreme end of normal crying behaviour; harmless, but can be distressing for carers.

Affects 20% of babies; m=f; it is pyloric stenosis that is classically seen in first born males!

Cause unknown, no proven treatment, but research shows that babies with colic continue to eat and gain weight normally. No medical treatment needed or proven, but do take a full history, undertake a thorough examination, and be supportive to parents.

Colic can cause much stress to unsupported individuals with poor coping strategies and the child may be at risk of abuse: helpful contacts: Cry-sis (www.cry-sis.org.uk) and National Childbirth Trust (www.nct.org.uk).

21. Answer is E.

SLE is a rare autoimmune disease with variable presentation and multi-system involvement: arthritis, photo sensitivity, butterfly facial rash, fibrosing alveolitis, pneumonitis, glomerulonephritis, pericarditis, cranial nerve lesions and anaemia being just some of its myriad presentations. 95% of patients are ANA (anti-nuclear antibody positive) on autoimmune profile testing.

An easy way to remember activators is UV PRISM:

UV – sunlight

Pregnancy
Reducing steroids
Infection – viral, bacterial
Stress
More drugs

Steroids are the mainstay of treatment; isoniazid, procainamide, hydralazine and some antibiotics have been implicated in drug-induced SLE.

22. Answer is A.

Also known as irritable hip – cause unknown; peak age 2–10 years; m>f; exclude septic arthritis; usually resolves spontaneously after 1–2 weeks.

23. Answer is D.

Perthe's disease – due to avascular necrosis of the femoral head; bilateral in 10%; peak age 4–7 years; m>f. Refer to orthopaedics for X-ray, rest, bracing, +/– surgery.

24. Answer is B.

SUFE – typically overweight underdeveloped children, or tall thin boys, aged 10–15 years; m>f; X-rays show backwards and downwards slip of femoral epiphysis with respect to femoral head; refer to orthopaedics for surgical pinning / reconstructive surgery.

25. Answer is C.

Septic arthritis of any joint is an emergency – child is very unwell; needs to be admitted for i.v. antibiotics.

26. Answer is A.

Amourosis fugax is caused by retinal emboli from ipsilateral carotid disease causing a temporary interruption to retinal circulation.

27. Answer is D.

The cause of the mature student's back pain and stiffness was ankylosing spondylitis, a common association of which is anterior uveitis.

28. Answer is E.

Herpes zoster ophthalmicus involves the tissues innervated by the ophthalmic division of the trigeminal nerve and accounts for 10–25% of all cases of shingles.

29. Answer is C.

Retinal vein occlusion can be of a branch or a central vein; as well as hyperlipidaemia, it is also associated with hypertension and hyperviscosity.

30. Answer is F.

Fluorescein staining of a dendritic ulcer caused by herpes simplex will show up when viewed through a blue light – treatment is with acyclovir eye drops – never use steroids as massive amoeboid ulceration and blindness can result.

31. Answer is H.

The painless loss of vision may be preceded by a storm of red floaters; one is unable to visualize the retina because of the haemorrhage within the orbit.

32. Answer is G.

Optic neuritis can be a presenting symptom of multiple sclerosis.

33. Answer is A.

Also known as Guillain–Barre syndrome. Hansen's disease is leprosy.

34. Answer is A.

Her best eye opening response is 3, her best verbal response is 3 and her best motor response is 6. Therefore, the maximum possible GCS score is 15.

35. Answer is D.

The correct date is 14/9/1998 (*BNF 54*, Sept 2007).

Questions 36–39: All answers taken from *Oxford Handbook of General Practice* (2005)

36. Answer is C.

Normal ear temperature ranges from 35.8 to 38.0°C.

37. Answer is B.

Normal oral temperature range is 35.5 to 37.5°C.

38. Answer is A.

Normal axillary temperature is 34.7 to 37.3°C.

39. Answer is D.

Normal rectal temperature is 36.6 to 38.0°C.

40. Answer D is true.

CBT and self help should be considered first-line in management of mild depression, rather than antidepressants because the risk–benefit ratio is low; if an antidepressant is to be prescribed, then an SSRI is recommended; they are as effective as tricyclics and are more likely to be continued because they have fewer side effects; for those who have had a recent MI, sertraline is considered the best antidepressant choice; venlaflaxine should not be prescribed to those who have recently had an MI or are at high risk of arrhythmias (*NICE*, April 2007).

41. Answer is D.

Turbid urine, fever, tenderness and a dipstick positive for leukocyte esterase all point to infection; loin pain and being acutely unwell suggest pyelonephritis.

42. Answer is A.

Positive glucose indicates diabetes; positive ketones indicate insulin is lacking and that adipose tissue is being metabolised, typical of IDDM; in the absence of glucosuria, ketone bodies suggest starvation.

43. Answer is C.

The loosest skin in a child is peri-orbital, so this is the first place that oedema is often noticed; the most common cause of nephrotic syndrome in children is minimal change glomerulonephritis.

44. Answer is B.

Haematuria can be due to numerous causes including inflammation, trauma, glomerulonephritis, calculi, instrumentation, neoplasms, etc; however, the severe pain suggests calculus.

45. Answer is G.

The macroscopic appearance would suggest hematuria, but the dipstick for blood is negative; this could be as a result of food colouring.

46. Answer is E.

Haematuria with atypical uroepithelial cells in a heavy smoker should prompt an urgent referral to urologists (2 week rule); smoking, exposure to beta-naphthylamine and analine dyes (rubber, petrochemical, paint, textile, petroleum industries) are also risk factors.

47. Answer is A.

Baglin *et al.* (*British Journal of Haematology*, 2006; **132**: 227–85) quote two randomized trials comparing a target INR of 2.5 (range 2.0–3.0) to a target greater than 3.0 (3.1–4.0); based on these studies, both groups of authors (Crowther *et al.* 2003; and Finazzi *et al.* 2005), concluded that a target INR of 2.5 was sufficient for the treatment of patients with thrombosis (venous or arterial) in association with anti-phospholipid syndrome; there are insufficient data to make an evidence-based recommendation for patients with anti-phospholipid syndrome and arterial thrombosis, but a higher target of 3.5 is often used.

48. Answer is A.

Target INR is 2.5 – grade A recommendation as per Baglin *et al.* (*British Journal of Haematology*, 2006; **132**: 227–85).

49. Answer is A.

Target INR is 2.5 – grade A recommendation as per Baglin *et al.* (*British Journal of Haematology*, 2006; **132**: 227–85).

50. Answer is C.

The newer second generation heart valves have a target INR of 3.0 (2.5–3.5) .

Questions 51–100

for answers see pages 31–36

51. Abdominal pain

Abdominal pain in a 4 year old can be the presenting feature of all the following except which one?

A Viral infection
B Appendicitis
C Meningitis
D Pneumonia
E Hypertrophic pyloric stenosis
F Migraine

52–55. Leg ulcers

A Neuropathic
B Arterial
C Neoplastic
D Venous

Match the possible leg ulcer diagnoses to the scenarios below. Each answer can be used only once.

52. A 75 year old Caucasian lady presents with an 18 month history of a shallow, intermittently healing leg ulcer, situated above the left medial malleolus. There is marked bilateral peripheral oedema.

53. An 80 year old man, known to be hypertensive and a heavy smoker, presents with a painful lesion at the tip of his left toe; he has been complaining of calf pain when he walks. The ulcer has a punched out appearance to it.

54. An obese 76 year old diabetic Asian lady attends with her son who is concerned about a foot infection; on examination, there is a deep painless ulcer over the head of the metatarsal of her left foot.

55. A 67 year old man with a long-standing venous ulcer has been seeing the practice nurse for treatment; she is concerned because the edge of the ulcer is not healing and the edges have started to evert.

56-66. Drug side-effects

A Glyceryl trinitrate
B Thyroxine
C Methotrexate
D Zopiclone
E Mesalazine
F Enalapril
G Micronor
H Rifampicin
I Atenolol
J Bendrofluazide
K Prednisolone
L Heparin
M Finasteride
N Eflornithine
O Oxytetracycline

Match the side-effects below to the drug most likely to cause them from the list above; match only one drug to each side-effect; not all the drugs will be matched. Note: there is more than one possible answer to some of the questions.

56. Adrenal suppression

57. Headache

58. Haemopoietic suppression

59. Gout

60. Exacerbation of Raynaud's phenomenon

61. Metallic taste in mouth

62. Dry cough

63. Orange–red tears

64. Hair growth

65. Hair loss

66. Staining of growing bones and teeth

67. Adiposity

Which one of the following is the most generally accepted measure of general adiposity in adults?

A Body Mass Index (BMI) centile
B Waist to hip ratio
C BMI
D Bioimpedance
E BMI *z*-score

68. Bariatric surgey

Which one of the following is not a mandatory criterion for referring an adult for bariatric surgery?

A The person commits to the need for long-term follow up
B The person has received or will be receiving intensive management in a specialist obesity service
C The person has a BMI of 40 kg/m^2 or more, or between 35 kg/m^2 and 40kg/m^2 and other significant disease (for example, type 2 diabetes or high blood pressure) that could be improved if they lost weight
D The person is generally fit for anaesthesia or surgery
E All appropriate non-surgical measures have been tried but have failed to achieve or maintain adequate, clinically beneficial weight loss for at least 12 months

69. Physical activity

What are the current levels of physical activity recommended by the Chief Medical Officer (CMO) for adults to achieve general health benefit? Choose only one answer.

A At least 30 minutes a day of at least moderate intensity physical activity on 5 or more days of the week
B At least 20 minutes a day of at least moderate intensity physical activity on 5 or more days of the week
C At least 30 minutes a day of at least moderate intensity physical activity on all 7 days of the week
D At least 10 minutes a day of low intensity physical activity on 5 or more days of the week
E At least 60 minutes a day of at least moderate intensity physical activity on 3 or more days of the week

70-72. Anaemia

Given normal values as follows:

- Hb 13.0–17.0 g/l (male)
- WCC 4.0–11.0 $\times 10^9$/l
- platelets 150–400 $\times 10^9$/l
- MCV 80–100 fl

A   Iron deficiency
B   Alcoholic liver disease
C   Aplastic anaemia
D   Sickle cell anaemia
E   Autoimmune hemolytic anaemia

For each of the results below, select the single most likely diagnosis from the list of options above.

70. Hb 8.2, WCC 6.3, platelets 246, MCV 110

71. Hb 5.1, WCC 0.4, platelets 34, MCV 84

72. Hb 9.4, WCC 7.9, platelets 175, MCV 76

73. Breast cancer

Risk factors for breast cancer include all except which one of the following:

A   High saturated fat intake
B   Unopposed oestrogen therapy
C   Breast feeding
D   Early menarche
E   Nulliparity

74. Health checks

General Practices must offer routine health checks to which one of the following groups:

A   Newly registered patients within 6 months of registering

B   Newly registered patients within 12 months of registering

C   Patients aged 16–75 who have not been seen by a health professional within the practice in the past 12 months

D   Patients aged >65 who have not been seen by a health professional within the past year

E   Temporary residents who are present for >2 months

75–79. Chest pain

A   Myocardial infarction

B   Pleurisy

C   Hyperventilation

D   Pneumothorax

E   Herpes zoster

F   Herpes simplex

G   Post-herpetic neuralgia

H   Fractured rib

Match one diagnosis above to each scenario below.

75. An 18 year old lady presents with difficulty breathing, chest tightness, and peri oral parasthesia; her heart sounds, breath sounds, bp and peak flow are all normal; she is tachypnoeic and is gasping for breath.

76. A 58 year old hypertensive man presents with sudden onset crushing central chest pain radiating to the jaw and left shoulder; he is pale, sweaty and vomits once.

77. A tall slim 21 year old male presents with sudden onset left-sided chest pain; he is dyspnoeic; the percussion note is resonant and breath sounds are reduced on the left; the trachea has been pushed over to the right.

78. A 78 year old lady in a nursing home presents with pain over the right side of her chest; she was seen by your colleague 2 days previously who thought the pain was musculoskeletal in origin; on examination she has a very painful vesicular rash affecting the T8 dermatome.

79. The previous lady is seen again a few weeks later; the rash has disappeared but the area remains exquisitely sensitive.

80. Prescription charges

Form FW8 needs to be completed to entitle which one of the following groups to free prescriptions?

A War pensioners
B Pregnant women and those with a child born less than 12 months ago
C Diet-controlled diabetics
D Insulin-controlled diabetics
E Those with a permanent fistula

81–85. Antibiotic treatment

A Flucloxacillin
B Penicillin V
C Co-amoxiclav
D Benzylpenicillin
E Amoxicillin
F Methicillin
G Penicillin and flucloxacillin
H None of these

Match the scenarios given below to the single most appropriate antibiotic treatment; each response may be used once, more than once or not at all.

81. A school teacher attends having been bitten the previous day by the class pet hamster; she has a small bite on the back of the left hand which looks inflamed; she is otherwise well.

82. A 1 year old child with headache, dislike of bright light and leg pains; on examination he has neck stiffness and a petechial rash.

83. A 15 year old girl complaining of a sore throat, fatigue and pyrexia for 10 days; examination shows sore throat, enlarged cervical and occipital lymph nodes; positive monospot blood test result received from lab.

84. A 37 year old checkout assistant presents with fever, malaise, cough productive of purulent sputum and right-sided pleuritic chest pain; coarse crackles can be heard at the right side of the chest.

85. A 4 year old child with red patch on cheek, covered in a golden crust. Child is otherwise well.

86-87. Genetics

A   HLA DR2
B   HLA DR3
C   HLA DR4

Match the disease to the genetic marker; there is only one correct answer; each response may be used once, more than once or not at all.

86. Addison's disease
87. Graves's disease

88. Sports medicine – banned substances

All of the following substances are banned in sports except which one?

A   anabolic steroids
B   human growth hormone
C   heroin
D   amphetamines
E   caffeine
F   diuretics

89-90. Classes of illegal drugs

A   Class A
B   Class B
C   Class C

Match the class to the drugs below (based on the Misuse of Drugs Act 1971).

89. Cannabis, benzodiazepenes, anabolic steroids.

90. Cocaine, LSD.

91. Contraception

Considering long-acting contraception in the form of Depo Provera, which one of the following statements is true?

A   Depo should automatically be used as first-line contraception for adolescents and teenagers

B   Depo may be used in women with risk factors for osteoporosis

C   Depo causes a reduction in BMI in most women

D   In women of all ages, one should carefully evaluate the pros and cons if the woman plans to use Depo for more than 2 years

92-98. Sick certification

A   Med 3
B   Med 4
C   Med 5
D   Med 6
E   MatB1
F   RM7
G   DS1500
H   SC1
I   SC2

Match one certificate to each situation below.

92. You are in doubt regarding the incapacity of patient; sent to DWP requesting early review of patient.

93. Patient has been getting regular sick notes, but has now received letter requesting statement prior to personal capability assessment.

94. Patient requests a back-dated sick note – was seen by you last week.

95. Mum-to-be in 21st week of pregnancy requires a certificate to allow her to claim statutory maternity pay.

96. Family attend requesting medical certificate – father is terminally ill and likely to die within 6 months.

97. First seven days of illness for self-employed plumber; not entitled to statutory sick pay.

98. First seven days of illness for employee of plumber; entitled to statutory sick pay.

99. Irritable bowel syndrome

Which one of the following list is not one of the Manning Criteria for diagnosis of IBS?

A   Abdominal pain
B   Increased stool frequency with pain
C   Looser stool with pain
D   Feeling of incomplete evacuation
E   Mucus in stool
F   Blood in stool
G   Relief of pain with defecation

100. Back pain

Which one of the following statements regarding the management of acute back pain is false?

A   NSAIDs are no better than placebo for acute pain
B   Muscle relaxants relieve pain more than placebo
C   There is strong evidence to suggest that bed-rest and specific exercises are not effective
D   Spinal manipulation may be considered for pain relief
E   Concordance with advice to stay active decreases chronic disability

Answers to questions 51–100

51. Answer E is untrue.
Abdominal pain in a four year old can be due to a dozen different reasons, not all of them intra-abdominal! Pyloric stenosis, however, presents with projectile vomiting, typically around the sixth week of life.

52. Answer is D.
Venous ulcers (which account for 70% of leg ulcers) are associated with a history of DVT, obesity, varicose veins, smoking; they have a shallow sloping edge and are often surrounded by an area of lipodermatosclerosis and varicose eczema; patients should have an ankle brachial pressure index done and if this is >0.8 and there are no contraindications, multi-layered, elastic, high compression bandaging should be applied.

53. Answer is B.
Arterial ulcers (which account for 10% of leg ulcers) are more painful when the legs are elevated but gravity helps increase blood flow when legs are dependent with consequent dependent rubour; unlike venous ulcers, they are usually dry. Associated gangrenous toes may be seen.

54. Answer is A.
Diabetic patients are at increased risk of arterial ulcers; however, their sensory perception is impaired because of neuropathy and so they are at risk of pressure from ill-fitting shoes – hence need for foot check as part of diabetes review in general practice.

55. Answer is C.
SIGN Guidelines (1998) state that if an ulcer is not healing after 12/52 or is showing suspicious changes such as a raised, rolled or everted edge, the patient should be referred for biopsy as the ulcer may be undergoing malignant change (Marjolin's ulcer). Also look at *NICE Guidance CG10 – Type 2 diabetes: Prevention and management of foot problems* (Jan 2004).

Questions 56–59: All answers taken from *BNF* (September 2007)

56. Answer is K.
Long-term steroids can also cause cateracts and osteoporosis.

57. Answer is A, but E is also a possibility (though less likely).
Headache caused by GTN tablet can be alleviated by swallowing tablet as stomach acids will inactivate it – this is why it needs to be taken sub-lingually, not orally.

58. Answer is C, but E is also a possibility (though less likely)

Patients need to have FBC, U/E and LFT before treatment, weekly for six months and then every 2–3 months; patients are advised not to self-medicate with aspirin or ibuprofen; alcohol should be avoided; all symptoms/signs of infection (especially sore throat) should be reported immediately.

59. Answer is J.

Thiazides can cause hyperuricaemia and gout.

60. Answer is I.

Beta blockers are contraindicated in severe peripheral vascular disease.

61. Answer is D.

Taste disturbance is associated with metronidazole, amiodarone, sulphasalazine, metformin and zopiclone – the last two typically leaving a metallic taste in the mouth.

62. Answer is F.

ACE inhibitors inhibit the breakdown of bradykinin and other kinins; angiotensin–II receptor antagonists, however, do not and so may be prescribed in those for whom the cough is troublesome.

63. Answer is H.

Urine, saliva and other body secretions are coloured orange–red by rifampicin and, unless warned, patients may be alarmed.

64. Answer is M, but G is also a possibility (though less likely).

Finasteride as 5 mg daily is prescribable on the NHS for BPH; as a treatment for male pattern baldness (1 mg o.d.) it is not available on the NHS.

65. Answer is N, but E and F are also possibilities (though less likely).

Eflornithine (Vaniqa) is a topical agent used to treat facial hirsutism in women; the two agents currently available to treat hair loss are finasteride in men and minoxidil in men and women.

66. Answer is O.

Deposition of tetracyclines in growing bones and teeth causes staining and occasionally dental hypoplasia; they should not be given to children under 12 or to pregnant/breast-feeding women.

67. Answer C is true.

According to *NICE Guidance CG43* (December 2006), BMI is the most widely accepted measure of *general* adiposity in the adult population. Adults with a BMI of 25 kg/m^2 are defined as overweight and those with a BMI of over 30 kg/m^2 are defined as obese.

Waist circumference is a useful measure of *central* adiposity in adults. Men with a waist circumference of 94 cm or more are at increased risk of health problems. If their waist circumference is 102 cm or more, even at a healthy weight (BMI 18.5–25 kg/m²) they are at increased risk. Women with a waist circumference of 80 cm or more are at increased risk of health problems. If their waist circumference is 88 cm or more, even at a healthy weight (BMI 18.5–25 kg/m²) they are at increased risk.

Waist-to-hip ratio is a useful measure of central adiposity in adults, but is more difficult to measure. There is no evidence on the utility of bioimpedance compared with BMI in adults.

BMI centile cut-offs are used as a measure of adiposity and adiposity change in children.

68. Answer E is untrue.

The time period for achieving or maintaining adequate, clinically beneficial weight loss prior to referral for surgery is 6 months. However, all the other criteria must be fulfilled.

Bariatric surgery is also recommended as a first-line option (instead of lifestyle interventions or drug treatment) for adults with a BMI of more than 50 kg/m² in whom surgical intervention is considered appropriate. Surgery for obesity should only be undertaken by a multidisciplinary team after a comprehensive preoperative assessment of any psychological or clinical factors that may affect adherence to post-operative requirements.

69. Answer A is true.

The Chief Medical Officer's 2004 recommendations (At least five a week: evidence of the impact of physical activity and its relationship to health, April 2004, DoH) emphasize that physical activity improves health from childhood to older age.

The recommended levels of activity can be achieved either by doing all the daily activity in one session or shorter bouts of at least 10 minutes. Activity can be structured sport/exercise, 'lifestyle activity' (climbing stairs, brisk walking, etc.) or a combination of these. These recommendations also apply to older people. Daily activity helps retain mobility, strength, co-ordination and balance.

It is likely that for many people 45–60 minutes of moderate physical activity per day is needed to prevent obesity. For good bone health, activity that produces high physical stresses on bones is necessary.

70. Answer is B, but E is also a possibility (though less likely).

Macrocytic anaemias are also found in hypothyroidism. Could also be E if there was a marked reticulocytosis pushing up the MCV.

71. Answer is C.

All parameters are low, this is aplastic anaemia.

72. Answer is A.

Iron deficiency presents as a hypochromic, microcytic picture.

73. Answer C is untrue.

Breast cancer is the commonest cause of cancer deaths in women, accounting for 18% of all female cancer deaths; British women have a 1:9 lifetime risk of developing this disease.

It increases with age; risk factors include nulliparity, early menarche, late menopause, oestrogen therapy unopposed by progesterone, positive family history and saturated fat intake; early first child and breast-feeding are protective.

74. Answer A is true.

Patients aged 16–75 who have not been seen within the past 3 years, and those aged >75 who have not been seen in the past year, should be offered a health check.

75. Answer is C.

76. Answer is A.

77. Answer is D.

78. Answer is E.

79. Answer is G

80. Answer is B.

Women should fill in form FW8 as soon as pregnancy is confirmed – the exemption certificate lasts for one year from the EDD; it may be completed after delivery, when it lasts 1 year from date of birth of child; it also covers dental charges.

81. Answer is C.

Co-amoxiclav is suggested by *BNF* for animal bites.

82. Answer is D.

Benzyl penicillin should be administered i.m. and the child immediately transferred to hospital as they may have meningococcal meningitis.

83. Answer is H.

None of the above as the girl has glandular fever; in particular, she should not be given amoxicillin which can cause a rash. Symptomatic treatment is advised.

84. Answer is E.
Amoxicillin, because this is a community-acquired pneumonia.

85. Answer is A.
Flucloxacillin for impetigo

86. Answer is B.

87. Answer is B.
Addison's disease and Graves's disease are both associated with HLA DR3. HLA DR4 is associated with rheumatoid arthritis. HLA DR2 may be associated with osteoarthiritis and with multiple sclerosis.

88. Answer is E.
Caffeine is allowed within permitted blood levels. Diuretics are useful to maintain fighting weight; frusemide can be used as a masking agent of other drugs in the urine.

Various drugs used to treat medical conditions, e.g. steroids, beta-blockers, insulin, clomiphene, and tamoxifen, are also subject to restrictions; in such cases, the athlete has to submit a TUE (therapeutic use exemption form) and prove that they have the medical condition mentioned *and* that they need to take medication for it.

Cold/cough remedies include banned stimulants such as ephedrine and pseudo-ephedrine.

The only medications known to be OK are paracetamol and ibuprofen.

89. Answer is C.
Cannabis has recently been down-graded.

90. Answer is A.
Heroin, methadone and ecstasy are all Class A drugs.

91. Answer D is true.
It is known that DMPA can reduce bone mineral density (BMD) and recent studies have confirmed this; the CSM therefore issued a warning to this effect in Nov 2004 (www.mhra.gov.uk); the reduction in BMD is constant, then plateaus after a few years; it is not known to what extent the BMD recovers, and this may be particularly important in adolescents who have yet to reach full bone mass; for this reason, although Depo can be used, it should only be used when other methods are unacceptable or unsuitable.

92. Answer is F.

93. Answer is B.

94. Answer is C.

95. Answer is E.

96. Answer is G.

97. Answer is H.

98. Answer is I.

Notes

- with Med 3, this can only be issued for date seen or one day after; only one Med 3 may be issued per person per period of sickness; may not be re-issued if lost; if needs another then mark as duplicate.
- if it more than one day since patient was seen then a Med 5 is issued; this is also issued if you have not seen patient, but sick note is being issued based on written report from another doctor; cannot forward date Med 5 more than one month.

99. Answer is F.

All of the others are in the Manning list of criteria, where if three or more are present, and there are no red flags, then a positive diagnosis of IBS can be made; blood in stool, weight loss, fever, anaemia or new symptoms in anyone over the age of 50 are all red flags and a full GI work-up is needed (*BMJ*, 2005; **330**: 632).

100. Answer A is false.

NSAIDs are better than placebo for acute back pain; the *BMJ* (2006; **332**: 1430) states that regular paracetamol and NSAIDs should be used; muscle relaxants and short term opioids may be considered, and also spinal manipulation.

Questions 101–150

for answers see pages 47–55

101. Secondary diabetes mellitus

The following are all associated with secondary diabetes mellitus, except which one?

A Thiazide diuretic therapy
B Haemochromatosis
C Primary hypoaldosteronism
D Pancreatic carcinoma
E Long term steroid use

102. Non-accidental injury

Which one of the following is more likely to be an accidental than non-accidental injury?

A Spiral fracture in the long bone of an infant
B Sub-dural haematoma in a baby
C Bruising on shins of a schoolboy
D Tear of frenulum (central fold behind upper lip)
E Multiple bruises of different ages

103. Haematology

The MCV (mean cell volume) is usually normal in which one of the following?

A Iron deficiency anaemia
B Chronic renal failure
C Folate deficiency
D Pernicious anaemia

104. Anaphylaxis

Which one of the following is not true? Anaphylaxis:

A is mediated by IgE antibodies, which cause release of histamine and other vasoactive mediators
B can be caused antibiotics, NSAIDs, vaccines and blood
C should be treated in the first instance with i.v. fluids such as sodium chloride
D can be fatal

105. Anxiety

In the non-urgent treatment of generalised anxiety disorder in primary care which one of the following is true?

A   Self-help is not a recommended treatment option

B   The optimal time range for Cognitive Behavioural Therapy (CBT) should be 40–50 hours in total

C   Monitoring of the efficacy and side effects of pharmacological therapy should take place within 2 weeks of starting treatment and again at 4, 6 and 12 weeks

D   If the first intervention tried does not produce an improvement after 12 weeks patients should be referred to a specialist mental health team

E   Serotonin-specific re-uptake inhibitors (SSRIs) are not first-line pharmacological therapy

106. Pandemic flu

With regard to pandemic influenza which one of the following statements is untrue?

A   Antigenic drift changes are usually the cause of pandemic strains of the influenza virus

B   Pandemics of human influenza can take place in any season

C   All previous human influenza pandemics have been caused by influenza A virus

D   The efficacy of antiviral drugs will only emerge once a pandemic is underway

E   It is estimated that, in the event of a pandemic, 14.5 million people will become ill

107. Diabetic retinopathy

In the management of diabetic retinopathy, which one of the following does not require urgent referral (i.e. patient to be seen in seven days or less) to an ophthalmology specialist?

A   Rubeosis iridis is present

B   There is evidence of new vessel formation

C   There is sudden loss of vision

D   There are hard exudates within 1 disc diameter of the fovea

E   There is evidence of retinal detachment

108. Childhood asthma

A 4 year old girl presents with persistent nocturnal and exercise-induced cough; she has been seen by your practice nurse who over the past few weeks has increased her medication and she is now on 2 puffs of salbutamol four times a day and 2 puffs twice a day of beclomethasone diproprionate; chest examination is normal and inhaler technique with a spacer is good.

Which of the following options is the next most appropriate step?

A   Trial of oral steroids

B   Double dose short-acting beta agonist

C   Leukotriene receptor antagonist

D   Measure peak flow

E   Admit to hospital

109-113. Headache

A   Sub-arachnoid hemorrhage
B   Cluster headache
C   Trigeminal neuralgia
D   Common migraine
E   Ramsay Hunt Syndrome
F   Giant cell arteritis
G   Sub-dural hematoma

Match one of the conditions above to each of the scenarios below.

109. 56 year old woman complains of sudden severe electric-shock like episodes of pain affecting the left lower jaw – pain lasts 1–2 minutes each time then fades to a dull throb for a few hours; pain is provoked by laughing or chewing; mother had something similar.

110. 30 year old man who smokes 20 cigarettes a day presents with severe right-sided headache centred around right eye which is red and watery; pain lasts for an hour at time and is waking him from his sleep – has been going on for two days now; had a similar episode lasting a week approximately six months ago affecting the same side and has remained symptom-free until now. He thinks it may be related to drinking red wine.

111. A 28 year old mother attends complaining of recurrent episodes of one-sided throbbing headache; these occur every four weeks or so and are accompanied by nausea, vomiting and photophobia; she has had them since her teens but they are becoming worse now.

112. An 80 year old woman attends complaining of headache and scalp tenderness and blurred vision in her right eye; she denies jaw claudication or systemic upset; on examination her temporal arteries are pulsatile and tender; her ESR is raised.

113. A 38 year old man attends complaining of severe occipital headache – he describes feeling as though someone has hit him on the back of the head with a baseball bat; a few days ago he had a slight headache which he thought odd as he has never had a headache before.

114. Alcohol intake

Current guidelines suggest that men should drink no more than how much alcohol per week (choose one answer only)?
A   14 units
B   21 units
C   28 units
D   14 pints
E   21 pints
F   28 pints

115. Coeliac disease

With regard to celiac disease, which one of the following statements is true?
A   The prevalence of coeliac disease in international population studies is 20%
B   Tissue transglutaminase antibody, endomysial antibody and immunoglobulin A should be used for initial testing
C   Unless antibodies are positive, coeliac disease cannot be diagnosed
D   Treatment involves a low carbohydrate diet, e.g. Atkins
E   Patients should be advised to start a gluten-free diet before any investigations are done
F   Patients should not be referred for further investigations unless antibody levels come back confirming coeliac disease

116. Gluten free diet

Which one of the following products is gluten-free?
A   Rye
B   Semolina
C   Barley
D   Rice

117-122. Glycaemic index

Rate the GI index of the following foods as:
A   high
B   medium
C   low

117. White rice.

118. Sweetcorn.

119. Lucozade.

120. Cherries.

121. Pear.

122. Banana.

123. Diagnosing diabetes

A patient is diabetic if, on testing they have (choose one of the following options):
A   a random blood sugar greater than or equal to 7.0 mmol/l
B   a fasting blood sugar greater than or equal to 7.0 mmol/l
C   a fasting blood glucose greater than or equal to 6.1 mmol/l
D   a blood sugar level of greater than 7.8 mmol/l after a formal glucose tolerance test; i.e. an overnight fast, drinking 75 g glucose (e.g. as 350 ml lucozade) and then having a blood sample taken 2 hours later

124-127. Vaginal discharge

A   Herpes
B   Candida
C   Trichomoniasis
D   Bacterial vaginosis
E   Physiological
F   Foreign body

Match the scenarios below to the diagnoses above – choose one in each case.

124. An Asian lady comes to see you, very distressed as she has developed a nasty vaginal discharge; she describes it as being very watery and grey. It is interfering with her ritual ablutions.

125. A young female student attends surgery complaining of a copious, mucopurulent, offensive, frothy green discharge; she thinks she may have a water infection as it is very sore to pass urine.

126. A heavily pregnant lady attends surgery worried about a heavy vaginal discharge; she describes it as having no odour, and as being thick and creamy, like cottage cheese.

127. A 26 year old married lady attends surgery with high fever and myalgia; she has multiple painful sores on the perineal mucosa and is finding it very difficult to pass urine because of the pain.

128-131. Aromatherapy

A   Indigestion
B   Insomnia, anxiety
C   Burns, relaxation
D   Clearing blocked noses

Match the aromatherapy oil below with its common use (choose only one in each case):

128. Lavender.

129. Valerian.

130. Peppermint.

131. Eucalyptus.

132. Hypokalaemia

Causes of hypokalaemia include all except which one of the following?
A   Long-term steroid use
B   Cushing's syndrome
C   Conn's syndrome
D   Addison's disease
E   Intestinal fistula

133. Hyponatraemia

Causes of hyponatraemia include all of the following except which one?
A   Diabetes insipidus
B   Addison's disease
C   Polydipsia
D   Diarrhoea / vomiting
E   Renal failure

134–137. Gynaecomastia

A Cancer bronchus
B Thyrotoxicosis
C Physiological
D Drug-related
E Liver disease
F Kleinfelter's syndrome
G Testicular tumours

For each case presented below, choose one from the list of common causes of gynaecomastia above.

134. A 26 year old man is attending, worried about breast swelling and tenderness; on examination the breast area is enlarged and slightly tender; he is also tender over his epigastrium and when questioned admits to taking his father's cimetidine, which he has found helpful in relieving an ongoing problem with indigestion.

135. A 37 year old teacher attends surgery; he is very anxious as he has developed enlarged swollen breast tissue; he feels agitated all the time but assumes he is worried about an imminent school inspection and is not sleeping; he has also lost 7 kg in weight over the past 2 months; on examination you note he is clammy, his pulse is 106/minute and his blood pressure 140/80; he is currently taking bendrofluazide 2.5 mg once daily for hypertension.

136. A 20 year old student attends surgery having developed swelling affecting the breast area; he is otherwise fit and well; he drinks 28 units of alcohol per week and smokes 10–15 cigarettes per day; on examination you notice his left testis feels hard.

137. A 68 year old retired builder comes to see you as he is feeling generally unwell; he has lost 7 kg in weight over the past month and is generally off his food; on closer questioning he has had some chest discomfort but this is mainly localized to the breast area; he has had a tickly cough for a few weeks which is not settling; he has smoked 40/day since the age of 16 but recently has gone off them; he is on tablets for an ongoing problem with an enlarged prostate but cannot remember what they are called. On examination you note him to have clubbing of the fingers and there are decreased breath sounds on the left side; the percussion note is dull.

138. A young couple come to see you; they have been married for 2 years but have not managed to start a family yet; you note the husband is very tall; on examination he has painless breast enlargement and small testes.

139. A jolly 56 year old publican comes to see you; he has been getting some jokes from his customers for developing 'man boobs' and is wondering if he can be referred for surgery; he drinks and smokes with his customers but has no idea how much; on examination you notice multiple spider naevi over his chest and palpation of the abdomen reveals a sharply demarcated mass arising from under the right costal margin; notes reveal concerns regarding his LFTs.

140–146. DVLA and fitness to drive

A Refusal/revocation of licence
B 1 week off driving
C 4 weeks off driving
D 6 weeks off driving
E 11 months
F 12 months
G Permanently barred
H No driving restrictions

For each case presented, choose the correct length of time or other consequence in terms of driving licence from the choices above.

140. A 22 year old student attends surgery having recently been seen by the local neurologists for investigation of recurrent episodes of unconciousness; he has been told he has epilepsy and cannot drive; since being started on medication 1 month ago, he has been fit free; he would like to know for how much longer he cannot drive.

141. A 68 year old retired school-mistress attends surgery; she had a single short-lived episode of weakness affecting the right hand 2 weeks ago; at the time she was also noted to have difficulty speaking and was slurring her words; her husband states that the whole episode lasted less than 10 minutes and she has been perfectly well since; she takes enalapril for hypertension and smokes 5 cigarettes per day.

142. A 44 year old housewife is worried about driving her child to school after she has had a pacemaker inserted next week; how long will she be unable to drive?

143. A 44 year old lorry driver is wondering whether it is ok for him to continue driving after his pacemaker insertion next week; you advise him, 'No.' How long will he be unable to drive for?

144. A 38 year old bus driver is diagnosed as being HIV positive.

145. A 59 year old lady has had a myocardial infarction; her husband attends surgery and informs you she is to have a CABG later that day; he would like to know how long she is likely to need off driving.

146. A 40 year old man is colour blind and has a visual acuity of 6/9 in both eyes.

147. Antibiotic prophylaxis

Which one of the following scenarios requires antibiotic prophylaxis?

A A 50 year old lady with a past history of infective endocarditis having a cervical smear

B A 60 year old man with a prosthetic heart valve *in situ* having a blood test

C A 36 year old lady with past history of infective endocarditis having a routine IUCD insertion 6 weeks after a normal vaginal delivery

D A heavily pregnant, anaemic lady found to have a previously undiagnosed murmur, due to have a dental extraction

E A 22 year old female student who wishes to have a pinna and nipple piercing on the same date

148. Notifiable diseases

All of the following are notifiable under the Public Health (Control of Disease) Act 1984 except which one?

A Mumps

B Measles

C HIV

D Dysentery

E Tuberculosis

149. Head lice

The following are true of head lice except which one?

A They typically affect the scalp with a predeliction for the nape of the neck and behind the ears

B Key findings of a large European study concluded that being from a family of lower social class, having more siblings and longer hair, increased chances of having head lice

C Salicylate emulsion is helpful in their management

D Using a louse comb is more effective for detection of infestation than visual inspection alone

E Dimeticone treatment is considered effective

150. Glaucoma

The following are all important signs of glaucoma, except which one?

A Visual field constriction

B Increased intraocular pressure

C Proptosis

D Severe pain affecting eye

Answers to questions 101–150

---

101. Answer is C.

Hypoaldosteronism leads to raised blood pressure and low potassium; most are caused by an aldosterone-secreting adenoma.

102. Answer is C.

Spiral fractures are often the result of a twisting force being applied to a limb; the frenulum can be torn if a child has a dummy (pacifier) or bottle pushed aggressively into the mouth; bruises or injuries of different ages indicate that the abuse has been going on for some time, rather than being an isolated incident.

103. Answer is B.

The MCV is decreased in iron deficiency anaemia and increased in pernicious anaemia and folate deficiency.

104. Answer is C.

Anaphylaxis is treated with oxygen, adrenalin, antihistamine and then hydrocortisone and/or i.v. fluids.

105. Answer is C.

The NICE guidelines on the management of anxiety were published in December 2004. Self-help, psychological and pharmacological therapies are all options for primary care management.

Self-help can include group CBT, bibliotherapy, support groups, and encouraging patients to improve their general health.

CBT should be delivered by a trained, supervised specialist adhering to treatment protocols. The optimal range of CBT is 16–20 hours in total (in weekly sessions of 1–2 hours completed within 4 months). Brief CBT can be delivered over a shorter time frame.

Pharmacological therapy should be an SSRI as a first-line, unless otherwise indicated. A number of factors need to be considered when prescribing (age, previous treatments response, risk of self harm, etc.) and patients need verbal and written information on side effects, withdrawal, delay in onset, etc., at the time of prescribing. Monitoring should take place at 2, 4, 6 and 12 weeks after starting treatment and at 8–12 week intervals if the drug is used for more than 12 weeks.

All interventions should be assessed for effectiveness after a course of treatment (using short self-complete questionnaires to monitor outcomes wherever possible). If there is no improvement, referral to specialist mental health services is usually made after at least two interventions have been tried.

106. Answer is A.

Annual outbreaks of ordinary 'flu' arise as a result of minor genetic changes in the flu viruses – antigenic **drift** producing different strains of the virus annually.

Pandemic flu is a result of major genetic changes in the flu virus – antigenic **shift**. These shifts have occurred sporadically throughout history and result in a novel flu virus to which the population has little or no immunity.

Flu viruses are divided into three main groups: influenza A, B and C. The A viruses cause most 'ordinary' flu epidemics and all previous pandemics. Influenza B and C viruses infect humans only, but influenza A also infects birds and other animals such as pigs and horses. This ability to 'jump' species enables influenza A to cause pandemics.

Pandemics of flu occur sporadically in any season compared to 'ordinary' flu which occurs every year during the winter months in the UK.

The World Health Organization (WHO) has advised that contingency planning is based on a cumulative clinical attack rate of 25%. The cumulative clinical attack rate is the percentage of the total population that are infected with the virus and display clinical symptoms. The attack rate of 'ordinary' flu is 5–10%. Using these rates it is estimated for the UK that 14.5 million people will be ill, there will be 50 000 additional deaths and 14 million extra GP consultations during the outbreak. At the peak of the pandemic there will also be 20 000 extra hospital admissions per week.

Vaccination and antiviral drugs are the medical counter-measures to the virus but have limitations. Antivirals have not been tested in pandemic situations so their effectiveness (particularly their impact on reducing mortality) is unknown. It is also not clear which patient groups will benefit most from the drugs and whether the virus will develop resistance to them.

The Department of Health (UK) have purchased vaccines against A/H5N1 virus. These may offer protection against a pandemic strain if it is similar. If the pandemic strain differs from A/H5N1 they will not confer protection but may act as a springboard for more rapid development of an effective vaccine. It is highly unlikely that an effective vaccine will be available early on in the epidemic, it may not confer 100% protection, and there will be insufficient supplies for the whole population to be vaccinated simultaneously. High risk groups will need to be prioritised.

For further information see:
- Department of Health Pandemic Flu: www.dh.gov.uk/en/PandemicFlu/index.htm
- Health Protection Agency: www.hpa.org.uk/infections/topics_az/influenza/pandemic/default.htm
- Royal College of General Practitioners: www.rcgp.org.uk/guidance/pandemic_planning.aspx

107. Answer is D.

Diabetic retinopathy is the leading cause of blindness in people under 60 in industrialised nations. It is also a major cause of blindness in older people. Twenty years after the onset of type 2 diabetes, over 60% of sufferers will have diabetic retinopathy but many will be asymptomatic until the disease is very advanced. The risk of visual impairment and blindness is reduced by care that combines screening with effective treatment. Screening must identify those with sight-threatening retinopathy that requires immediate preventative treatment.

If there is a sudden loss of vision the NICE guidelines on the management of diabetic retinopathy (published in 2002) recommend that the patient be seen within 24 hours by an ophthalmology specialist. This is also the case if there is evidence of retinal detachment. If new vessel formation, rubeosis iridis, or pre-retinal and/or vitreous haemorrhage is detected, patients should be seen within one week. An unexplained drop in visual acuity, hard exudates within one disc diameter of the fovea, macular oedema, unexplained retinal findings and pre-proliferative (or more advanced) retinopathy should be seen by a specialist within a maximum of four weeks.

108. Answer is C.

Step 3 of the British Thoracic Society guidelines state that for a child on p.r.n. short-acting beta agonist and regular standard dose inhaled corticosteroid therapy, the next therapeutic step is to add on a leukotriene receptor antagonist.

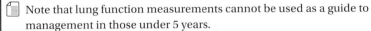

Note that lung function measurements cannot be used as a guide to management in those under 5 years.

109. Answer is C.

TN (also known as tic douloureux) – paroxysmal bursts of severe burning or shock-like pain lasting from a few seconds to up to two minutes at a time; typically felt on one side of the jaw or cheek; asymptomatic in-between attacks which can last from a few days up to months; triggers for attacks include chewing, laughing, shaving or being exposed to the wind; most often in people>50, f>m; can run in families perhaps because of an inherited pattern of blood vessel formation – ? an abnormal blood vessel pressing on trigeminal nerve. Treatment options include medication (anti-convulsants, tricyclics), complementary (acupuncture, biofeedback), and surgery; *NICE* (IPG085, August 2004) approved stereotactic surgery for TN using the gamma knife.

110. Answer is B.

The *Migraine in Primary Care Advisors Group* have issued treatment guidelines for cluster headaches (last reviewed in August 2006) recommending prophylaxis using steroids in the short term and verapamil as the gold standard for long-term prophylaxis; subcutaneous sumatriptan is the gold standard treatment to be used as a rescue medication when break-through attacks occur despite the use of prophylaxis (www.clusterheadaches.org.uk).

111. Answer is D.

The New Generalist (Volume 3, Autumn 2005) states that a stratified care approach where the patient receives the treatment most appropriate to their illness severity and frequency is preferable to the traditional step care analgesic approach: all patients will require acute medications (analgaesics +/- anti-emetic for mild/moderate, triptan for moderate/severe) in the first instance; prophylaxis may be required for those with frequent (>3–4 attacks/month), disabling attacks, or for those who find acute medications ineffective/intolerable; first-line prophylactic agent likely to be a beta-blocker in most cases (neuromodulators such as topiramate, amitriptyline, and calcium channel blockers also used but not all are licensed for this use in the UK).

Some complementary medications such as feverfew, magnesium, vitamin B2 (riboflavin), acupuncture and butterbur may be used in *addition to, but not instead of* usual medication (*MIPCA Guidelines*).

112. Answer is F.

Giant cell arteritis with neuro-ophthalmic complications requires a temporal artery biopsy as soon as possible to provide a tissue diagnosis to justify the long-term systemic steroid therapy (and its associated side effects) that this condition requires, and also to help exclude other diagnoses, e.g. stroke. Treatment with high dose steroids initially and gradually reducing can take between six months and a year, or longer (*Br J Ophthalmol*, 2001; **85**: 1248–51).

113. Answer is A.

Sub-arachnoid haemorrhage is usually due to a bleed from an aneurysm and accounts for 5% of all strokes; half of patients are <55 years old and outcome is generally poor: it has a high mortality (50% die within a month) and morbidity (of survivors, 50% become dependent in terms of help with ADL – activities of daily living).

114. Answer is B.

Current recommendations are no more than 14 units for women and 21 for men.

115 Answer B is true.

Coeliac disease prevalence is 0.5–1% in international studies. Antibody-negative coeliac disease with villous atrophy is now recognised as a disease entity and patients who remain symptomatic despite negative blood tests should be referred for further testing; patients in whom coeliac disease is suspected should avoid starting a gluten-free diet until diagnostic confirmation with duodenal biopsy.

116. Answer is D.
Rice, tapioca and sago are gluten free, but wheat, barley, semolina and rye all contain gluten.

117. Answer is A.

118. Answer is B.

119. Answer is A.

120. Answer is C.

121. Answer is C.

122. Answer is B.
The glycaemic index is a numerical system that tells you how fast a particular food triggers a rise in your blood sugar levels; the higher the GI, the faster that particular food will cause a rise in blood sugar; for people such as diabetics, eating lower GI foods is thought to be beneficial by allowing a steady release of sugar into the bloodstream, reducing the need for snacking and helping with weight loss.

123. Answer is B
A fasting level of 7.0 mmol/l or more, a random/post-GTT sample of 11.1 mmol/l or more, are diagnostic of diabetes mellitus.

Fasting samples of 6.1 mmol/l or more, but less than 7.0 mmol/l, are indicative of impaired fasting glycaemia.

A GTT sample of 7.8 mmol/l or more, up to 11.1 mmol/l, implies the patient has impaired glucose tolerance.

Both impaired fasting glycaemia and impaired glucose tolerance are risk factors for developing diabetes and the patient should be followed up with annual blood tests.

Questions 124–127. Answers drawn from *BMJ*, 2004; **328**: 1306–8.

124. Answer is D.
Ritual ablutions (or wudu) include washing of the perineal area with water; any discharge breaks the wudu, and soils the clothing, and so the patient has to perform the whole process again in order to enter a state of prayer. Treatment options include: metronidazole (oral, topical) or clindamycin (note:

intravaginal clindamycin can cause condom failure, and because it also kills lacto bacilli can predispose to vulvovaginal candidiasis). Note also that relapse is common; the bacteria responsible do not persist in male partner and so treatment of male partner does not affect relapse rate.

125. Answer is C.

Symptoms of infection with the flagellated protozoan *Trichomonas vaginalis*, can also include vulvar irritation and superficial dyspareunia; transmission is usually sexual; external genital examination may be normal (classic strawberry cervix appearance due to punctate haemorrhage is uncommon); treatment options include oral metronidazole or tinidazole; side effects of metronidazole include nausea, metallic taste in mouth, disulfiram reaction with alcohol so problems with compliance with lower dose and longer course regimens, although WHO prefers 5 day course for men, rather than single stat dose of 2 g; patient needs follow up for test of cure and contact tracing.

126. Answer is B.

Candida infection is common; affects 75% women during their reproductive life; is associated with diabetes, pregnancy (or both!), antibiotic use and immunosuppression; transmission is mostly non-sexual; patient may present with external dysuria, superficial dysparuenia, as well as classic cottage cheese discharge; a large number of oral and topical preparations are available. Note that miconazole and econazole have an adverse effect on latex condoms so can cause condom failure.

127. Answer is A.

Herpes is the leading cause of genital ulcer disease worldwide (*BMJ*, 2007; **334**: 1048–52); counselling of patient and partner is very important; presentation of first episode is often severe and associated with systemic upset; urinary symptoms can be severe enough to include urinary retention; attacks can be managed symptomatically and antivirals such as acyclovir are recommended: they do not cure the illness but can reduce viral shedding and severity of illness; patient should be advised to use barrier method and avoid sexual intercourse when symptomatic; oral sex should be avoided when a cold sore is present.

The patient or her partner may have been carrying the virus and been asymptomatic for now, so this current event need not necessarily be a sign of infidelity.

For further information see: www.herpes.org.uk.

128. Answer is C.

Lavender can be used to aid relaxation and to help relieve pain and assist healing of burns.

129. Answer is B.

Valerian is useful in insomnia (also available in tablet form for oral use).

130. Answer is A.

Peppermint is useful for indigestion; and is used in the oral form to help with indigestion and bloating.

131. Answer is D.

Eucalyptus oil is very useful for clearing congestion.

132. Answer is D.

Addison's disease causes hyperkalaemia.

Other causes of hypokalaemia include diarrhoea, vomiting, purgative abuse.

Causes of hyperkalaemia include renal failure, ACE inhibitors, potassium-sparing diuretics.

133. Answer is A.

Diabetes insipidus causes an increase in sodium levels.

134. Answer is D.

Cimetedine, spironolactone and verapamil are all associated with an increased risk of gynaecomastia; the anti-ulcer drugs ranitidine and misoprostol are not (*BNF*, Sept 2007).

135. Answer is B.

Although thyrotoxicosis is more common in women than men, it can occur in men where it can cause the classic symptoms as described, as well as gynaecomastia and sexual dysfunction.

136. Answer is G.

The incidence of gynaecomastia in adult men is reported as being 35–65% depending on criteria used; 2% of men presenting with gynaecomastia are found to have testicular tumours; it is important to do a testicular examination on all men presenting with breast enlargement and advise those with a normal examination to continue to self-check regularly and report any abnormalities to their GP (*BMJ*, 2006; **332**: 837–8).

137. Answer is A.

This man has lung cancer and a pleural effusion on the left side; the tumour is releasing chemicals that are causing gynaecomastia – this is known as a paraneoplastic syndrome.

138. Answer is F.

Men with Kleinfelter's syndrome may have breast enlargement; they also tend to have testicular atrophy and be azoospermic.

139. Answer is E.

This man has signs of liver failure secondary to chronic excess alcohol consumption.

140. Answer is E.

Note that he was started on medication 1 month ago and has been fit-free since; hence only 11 months to go!

141. Answer is C.

DVLA at a Glance Guide (page 8) states that a patient having a single TIA must not drive for at least one month and may resume after this time if the clinical recovery is satisfactory.

142. Answer is B.

Pacemaker implant (includes box change): driving must cease for at least one week; may be permitted thereafter provided there is no other disqualifying condition.

143. Answer is D.

Pacemaker implant disqualifies Group 2 driver for 6 weeks. Re-licencing may be permitted thereafter provided there is no other disqualifying condition.

144. Answer is H.

HIV is not a bar to driving either a group 1 or group 2 vehicle.

145. Answer is C.

CABG: driving must cease for 4 weeks – may resume thereafter provided no other disqualifying condition exists; DVLA do not need to be notified.

146. Answer is H.

Colour blindness is not a bar to driving a group 1 or group 2 vehicle; visual acuity must be satisfactory in both colour blind and normal-sighted persons equally.

147. Answer is C.

For gynaecological procedures, it is recommended that antibiotic prophylaxis is given to women with prosthetic valves or who have had endocarditis previously; in these circumstances an intravenous regimen is advised; in the absence of specific guidance the FFPRHC considers that such prophylaxis should be used for both insertion and removal (*NICE Long Acting Reversible Contraception*: CG30, Oct 2005).

> Note that on 9 November 2007, *NICE* began consultation on a guideline regarding prophylaxis against infective endocarditis and a key recommendation is that women undergoing obstetric or gynaecological procedures do not require antibiotic prophylaxis; the RCOG/FFPRHC have not changed their mind yet so it is probably best to wait for full clinical guideline to be published before changing practice).

Cervical smear does not require antibiotic prophylaxis, nor does routine phlebotomy or dental procedures.

The murmur in the pregnant lady is probably a flow murmur.

The guidelines for the prevention of endocarditis: report of the Working Party of the British Society for Antimicrobial Chemotherapy (Gould *et al.* (2006) *J Antimicrob Chem*) states that where the process involves non-infected skin incision but no mucosal breach (e.g. in the case of nipple or pinna piercings), antibiotic prophylaxis is not needed but adequate skin disinfection should be carried out prior to the procedure.

148. Answer is C.

Other notifiable diseases include: acute encephalitis, acute poliomyelitis, anthrax, cholera, diphtheria, dysentery, food poisoning, leptospirosis, leprosy, malaria, measles, meningitis, mumps, ophthalmia neonatorum, paratyphoid fever, plague, rabies, relapsing fever, rubella, scarlet fever, smallpox, tetanus, typhus fever, vital haemorrhagic fever, viral hepatitis, whooping cough, yellow fever.

149. Answer C is false.

Do remember that, when treating an entire family, it is a legal requirement to issue one prescription per person (*Drugs and Therapeutic Bulletin*, July 2007).

150. Answer is C.

Proptosis is a forward displacement of the orbit and is not a sign of glaucoma.

Questions 151–200

for answers see page 65–70

151–156. Herbal remedies

A   Indigestion
B   Migraine
C   Depression
D   Eczema
E   Prevention and treatment of common cold and other viruses
F   Insomnia

Match the herbal remedy below with its common use.

151. St John's Wort

152. Feverfew

153. Chinese herbal medicine

154. Echinacea

155. Peppermint

156. Valerian

157. Seborrhoeic warts

Which one of the following statements is true?
A   Arise from the sebaceous glands
B   Are common in people over the age of 50 years
C   Are found mostly on the hands and face
D   Are pre-malignant
E   Are caused by a wart virus

158–166. Haematology

A Felty's syndrome
B Burkitt's lymphoma
C Non-Hodgkin's lymphoma
D Iron deficiency anaemia
E Chronic myeloid leukaemia
F Acute lymphoblastic leukaemia
G Chronic lymphocytic leukaemia
H Multiple myeloma
I Vitamin B12 deficiency
J Vitamin D deficiency
K Sickle cell disease
L Hodgkin's lymphoma

For each scenario described below, select the single most likely diagnosis from the list above.

158. A 35 year old man presents with weight loss, lassitude and anaemia; on examination you note he has a markedly enlarged spleen; investigations reveal anaemia, raised white cell count, positive Philadelphia chromosome.

159. A 28 year old nurse attends with her fiancée who is a medical registrar; they had noticed a solitary enlarged rubbery cervical lymph node in the right sub-mandibular region; they are concerned as she appears to be losing weight for no obvious reason and have come in today for the results of blood tests; the haematologist has remarked on the presence of Reed–Sternberg cells.

160. A 7 year old Sudanese boy presents with night sweats, fever, weight loss; his mother has noticed a mass arising from the left lower jaw; blood tests show a positive monospot.

161. A 70 year old man is awaiting surgery for his prostate; routine blood testing shows him to have a raised lymphocyte count. The haematologist has commented on the presence of 'smear cells' and has already made an outpatient appointment to see him.

162. A 60 year old man presents with a 6 week history of progressive mid-thoracic back pain and tenderness; the main reason for attending today is that he has not been able to shake a recent chest infection and every time he coughs, the pain in his back is exacerbated; you organize some blood tests and when they return, they show anaemia, raised ESR and rouleaux formation; serum electrophoresis shows a paraprotein band and urine electrophoresis confirms the presence of Bence Jones proteins.

163. A 49 year old lady attends complaining of fatigue, a sore mouth and difficulty swallowing; she is also worried that she is losing hair and is wondering if you can give her anything for heavy periods.

164. A 36 year old vegan lady attends complaining of a sore mouth; you note her to have a smooth, sore tongue; she has numbness affecting her limbs in a 'glove and stocking' distribution; on examination her joint position and vibration sense are disturbed.

165. A 4 year old boy presents with fever, weight loss, and severe bilateral leg pains; he bruises easily and is suffering from recurrent chest infections; full blood count shows a pancytopenia.

166. A 7 year old boy of Afro–Caribbean descent presents with severe abdominal pain and swelling of the hands and feet; blood film shows presence of elongated crescent-shaped blood cells.

167. Disability living allowance

With regard to disability living allowance, which one of the following statements is true?

A Is means tested
B Is non contributory
C Has replaced attendance allowance for all age groups
D Has a mobility and a care component
E Is available for all age groups

168. Semen analysis

Which one of the following is false?

A Collection is by masturbation after abstinence from sexual activities, including masturbation for 3–5 days
B The sample is not tested for sperm antibodies
C If the first sample is abnormal, patients should be referred immediately with a view to sperm recovery and cryopreservation
D The sample must be taken to lab as soon as possible after production and be kept warm during transfer

169. Irritable bowel syndrome

Which one of the following is true regarding irritable bowel syndrome?

A   Typically presents in the over 50s

B   FBC, ESR or CRP are not helpful in diagnosis and should not be performed as they are not recommended to exclude other diagnostic possibilities

C   In management, insoluble fibre (bran) is more helpful than soluble fibre (ispaghula powder)

D   May respond to small doses of amitriptyline

E   Acupuncture and reflexology should be encouraged by primary care physicians as a means of self help

F   Is a diagnosis of exclusion using Duke's classification

170. Atopic eczema

Considering emollient therapy, which one of the following is true?

A   Bath emollients can cause allergic contact dermatitis

B   There is strong evidence to support the use of bath emollients over topical emollients

C   There is no general consensus amongst clinicians to confirm that the application of emollients directly to the skin is effective

D   The routine use of antiseptic/emollient preparations in patients with atopic eczema is highly recommended

171. Diabetes mellitus

Which one of the following statements is true?

A   Poor glycaemic control at conception and during pregnancy is associated with an increased risk of still birth

B   Symptoms alone (e.g. subjective feelings of weakness) are a very good indicator of biochemical hypoglycaemia (blood glucose <3 mmol/l)

C   In the case of a diabetic patient with COPD, extra blood glucose monitoring is not necessary during intercurrent infections, as long as any such exacerbations are treated with a course of oral prednisolone

D   NICE guidelines for patients with type II diabetes, suggest that self-monitoring of blood glucose levels can be considered a stand-alone intervention

172. Macular degeneration

Considering age-related macular degeneration, which one of the following is true?

A   NICE has stated that photodynamic therapy may be used to treat certain types of dry age-related macular degeneration

B   NICE has stated that photodynamic therapy may be used to treat wet age-related macular degeneration

C   NICE has approved the use of Lucentis (ranibizumab) in the treatment of dry age-related macular degeneration

D   NICE has not approved the use of Lucentis (ranibizumab) in the treatment of wet age-related macular degeneration

173. Impetigo

A 4 year old boy is brought in with a lesion on the right cheek, about the size of a two pence piece, which has been present for 2 days; it started off as an insect bite when he had been at his grandmothers; this then blistered, broke down, and the liquid dried to a golden brown crust; he is otherwise well and has had all his immunizations to date.

Which of the following statements should you make to the mother?
A   Current evidence shows that removing the crust and applying topical antiseptics is the treatment of choice and is highly effective
B   Not to worry, as the condition is not very contagious
C   His teachers are likely to be perfectly happy for him to continue to attend school
D   You need to contact social workers as this is a case of non-accidental injury.
E   The condition usually leaves permanent hyper-pigmentation and scarring
F   He may need antibiotics

174–178. Evidence-based medicine: studies

Consider the following types of study and put them in heirachical order of evidence – that is, put the study with the greatest weight as 174, the least as 178
A   Expert committee reports
B   High quality case-control study
C   Reasonably high quality randomised controlled trials
D   Cohort study with high risk of bias
E   High quality meta-analysis of randomised controlled trials

179–189. Statistical terms and studies

A Mode
B Mean
C Median
D Case-control
E Cross-sectional
F Cohort
G Confidence interval
H Attributable risk
I Relative risk
J Number needed to treat
K None of these
M Meta analysis

Match the definition to one of the options given above.

179. The sum of all the values given, divided by the number of values.

180. The mid-point of a given set of values such that half are below, half are above.

181. The most commonly occurring value.

182. A range, within which we can be fairly certain that the true value lies.

183. A descriptive study that provides a snapshot of the population being studied.

184. A prospective observational study that follows a group over a period of time and investigates the effect of a treatment or risk factor; can calculate incidence of a disease.

185. A retrospective study which investigates the relationship between a risk factor and one or more outcomes; carried out by selecting cases that already have disease, matching them to cases who are the same but disease-free, and then comparing the effect of the risk factor on the two groups. It looks for exposure and can be used to calculate the odds ratio.

186. Incidence of disease in exposed population divided by incidence of disease in non-exposed population.

187. Incidence of disease in exposed population minus incidence of disease in non-exposed population.

188. Reciprocal of attributable risk.

189. A method of combining results from a number of independent studies to give one overall estimate of benefit or harm in the forms of an odds ratio.

190–192. Cremation forms

A Form A
B Form B
C Form C

Match the form correctly to each scenario below.

190. Certificate of medical attendant to be signed by first doctor.

191. Application for cremation.

192. Signed by a second doctor who is not in partnership with the first doctor and who has been qualified for at least 5 years.

193. Chlamydia

The following are all suitable options for treating chlamydia except which one?
A Azithromycin 1 g stat.
B Doxycycline 100 mg b.d. 1 week
C Erythromycin 500 mg b.d. 2 weeks
D Topical clindamycin

194–198. Hypertension

A Angiotensin-converting enzyme inhibitor
B Calcium-channel blocker
C Thiazide diuretic
D Beta-blocker
E Angiotensin II receptor antagonists

Consider the following hypertensive patients, all of whom need treatment and, bearing recent evidence and guidelines in mind, suggest the single best class of anti-hypertensive for each patient from the list above.

194. A 40 year old Caucasian lady.

195. A 60 year old Caucasian man with a past history of gout.

196. A 40 year old man of Afro–Caribbean origin, also with a history of gout.

197. A 55 year old man who also suffers with angina.

198. A patient who was initially on enalapril but has a troublesome tickly cough.

199. Ischaemic heart disease mortality

Which combination of drugs has been shown to have the greatest reduction in all cause mortality in patients with a first diagnosis of ischemic heart disease?

A   Statin, aspirin, beta blocker

B   Statin, aspirin, ACE inhibitor

C   ACE inhibitor, beta blocker, statin

D   Statin, aspirin, folic acid

200. Confidence intervals

If a study quotes a 95% confidence interval, which one of the following statements is true?

A   There is a 95% chance of the true value lying outside these limits

B   There is a 5% chance of the true value lying outside these limits

C   There is a 2.5% chance of the true value lying outside these limits

D   There is a minus 5% chance of the true value lying outside these limits

Answers to questions 151–200

151. Answer is C.

A meta analysis of 23 RCTs shows St John's Wort is more effective than placebo and as effective as conventional anti-depressants in mild to moderate depression, with fewer side effects; significant interactions have been reported, especially with SSRIs, warfarin and COCP. There is now evidence that hypericum in doses of 900 mg is at least as effective as 20 mg paroxetine in moderate to severe depression and is better tolerated; physicians, however, are not encouraged to promote it but should be aware that many people do self prescribe OTC (*BMJ* 2005; **330**: 503–6).

152. Answer is B.

Both *Bandolier* and *Cochrane Database* (Pittler *et al.* 2000: Feverfew for preventing migraine) have stated that studies show that feverfew may be beneficial for the prevention of migraine; its effectiveness has not been established beyond reasonable doubt; adverse effects are minimal. Furthermore, the *Migraine in Primary Care Guidelines* have stated that complementary therapies such as feverfew, magnesium, vitamin B2, acupuncture and butterbur may be used in addition to, but not instead of, regular medication (www.mipca.org.uk/guidelines, 2006).

153. Answer is D.

Chinese herbal medicine is based on principles of yin and yang and aims to treat ill health through re-balancing these two elements and so allow the person's Qi to flow, thus restoring health; it has been used to treat a number of conditions, including skin disease.

154. Answer is E.

Echinacea, in a double-blind study, has been shown to be ineffective against the common cold (*NEJM*, 2005; **353**: 341–8). *Bandolier* and the *Cochrane Database* reported similar thoughts in 2000; however, many people do take this preparation OTC with or without zinc and vitamin C.

155. Answer is A.

Peppermint is used in many indigestion remedies both prescribed and OTC.

156. Answer is F.

Valerian is used in many stress and sleep aids (e.g. Kalms).

157. Answer is B.

Seborrhoeic warts are benign greasy-brown warty lesions, usually on back, chest, face and are very common in the elderly.

158. Answer is E.

CML is most commonly seen in middle age, often with a male preponderance; symptoms are chronic and insidious; those without the Philadelphia chromosome have a poorer prognosis.

159. Answer is L.

Lymphomas are a malignant proliferation of lymphocytes; Hodgkin's lymphomas are characterized by cells with mirror-image nuclei – the Reed-Sternberg cells; tend to occur as painless enlarged nodes.

160. Answer is B.

Burkitt lymphoma: endemic across certain regions of equatorial Africa; high-grade B cell neoplasm; endemic African form most often affects maxilla or mandible; sporadic (non-endemic) form tends to affect abdominal organs; Epstein–Barr virus has been implicated strongly in endemic form, relationship less clear with sporadic form; mean age in Africa: 7.

161. Answer is G.

25% of patients with CLL are asymptomatic; prognosis is good; affects elderly.

162. Answer is H.

Postural bone pain and tenderness is common in this neoplastic proliferation of plasma cells with diffuse bone marrow infiltration and focal osteolytic lesions; can present as pathological fracture; X-rays may show punched-out lesions (pepper pot skull). Multiple myeloma affects elderly.

163. Answer is D.

This lady has iron deficiency anaemia, which has led to angular cheilosis (fissures at angles of mouth), atrophic glossitis (smooth tongue) and the Plummer–Vinson (or Patterson–Kelly–Brown) syndrome of incoordinate movements in the pharynx, sometimes accompanied by actual web formation which causes dysphagia.

164. Answer is I.

This lady is deficient in vitamin B12 and has sub-acute combined degeneration of the spinal cord.

165. Answer is F.

Pancytopenia in a 4 year old boy is ALL.

166. Answer is K.

The crescent-shaped blood cells have 'sickled'; the child is suffering from hand and foot syndrome.

167. Answer is B.

Disability Living Allowance is payable to anyone who is under 65 years of age at the time of application who has had a disability lasting longer than 3/12 and where the disability is expected to last longer than 6 months; it has a mobility and a care component and is not means tested.

168. Answer is C.

NICE Guidelines (2004) state that if the first sample is abnormal, it should be repeated after 3 months to allow time for a cycle of spermatozoa formation to be completed.

Sperm antibodies are produced by the female; NICE states that screening for sperm antibodies should not be offered as there is no effective treatment to improve fertility.

169. Answer is D.

IBS typically presents in the 20–40 age group; Manning's criteria is helpful in diagnosis, which is one of exclusion (Duke's classification is used for staging of colorectal adenocarcinoma). Initial investigations like FBC, ESR, CRP may be helpful in excluding sinister causes for symptoms. *NICE Draft Guideline* (August 2007).

170. Answer is A.

Although emollient bath additives are often prescribed for the treatment of eczema and related disorders, the evidence base for their use is not strong (*Drugs and Therapeutics Bulletin*, October 2007).

171. Answer is A.

It is very important for women of child-bearing age who are diabetic to have prenatal counselling to avoid a poor perinatal outcome (*Drugs and Therapeutics Bulletin*, September 2007).

172. Answer is B.

Remember, Wet is Worse – although the rarer of the two (1 in 10 cases) and treatable, disease onset can be very rapid.

Dry ARMD (9 in 10 cases) has no treatment available (dietary supplements may help), but tends to be a much slower progressing disease and even after many years, many people do not totally lose their reading vision (*NICE Draft Guidelines*, September 2003 and June 2007).

Ranibizumab is a monoclonal antibody fragment with anti-angiogenic properties; it is injected monthly intravitreally (i.e. into the vitreous humour of the eye); it inhibits vascular endothelial growth factor A which is thought to trigger choroidal neovascularisation and macular oedema in wet ARMD.

173. Answer is F.

There is insufficient evidence to show whether removing crust or using topical antiseptics is likely to be helpful.

Topical fucidic acid for 7 days as a first-line treatment is a reasonable option; however, there is a growing problem with *Staph. aureus* resistance in the community in which case topical mupirocin may be considered, but only if the cause is MRSA.

Oral flucloxacillin may be needed if the infection is crusted and extensive, or bullous in nature. If streptococcal infection is known or suspected, phenoxymethylpenicillin can be added to flucloxacillin (*Drugs and Therapeutics Bulletin*, January 2007).

174. Answer is E.
The study with the greatest weight in terms of evidence based medicine is a high quality meta-analysis of randomised controlled trials.

175. Answer is C.
Next is a reasonably high quality RCT.

176. Answer is B.
Then a high quality case-control study.

177. Answer is D.
A cohort study with a high risk of bias is less good.

178. Answer is A.
An expert committee report is regarded as having the weakest evidence base.

179. Answer is B.
The mean of the numbers 5, 6, 6, 6, 7, 9, 10 is $(5+6+6+6+7+9+10)/7 = 49/7 = 7$.

180. Answer is C.
The median of the set of numbers above is 6.

181. Answer is A.
The mode (most frequently occurring number or event) of the same set of numbers is 6.

182. Answer is G.
If the confidence interval is 95%, we can be 95% certain that the true value lies within these limits.

183. Answer is E.
A cross-section at a point in time, e.g. undertaking a survey to see prevalence of chlamydia in a population.

184. Answer is F.

For example, a cohort study in the *BJGP* in 2004 looked at the fat intake in a group of newly diagnosed non-insulin dependent diabetics over a (prospective, moving forwards) 4 year period and found that, at diagnosis, many had an unfavourable fat intake, but that soon after, a majority adopted a more favourable consumption pattern and this improved pattern was maintained over the 4 year period.

185. Answer is D.

For example, looking back (retrospective), the use of thalidomide is compared between a group of women having abnormal babies and those having healthy ones.

186. Answer is I.

For example, if the risk of developing a DVT on pill A is 24/1000 and the risk on pill B was 8/1000, then the relative risk would be 0.024/0.008 = 3.

187. Answer is H.

For example, the incidence of disease X in smokers is 20% but the incidence in non-smokers is 2%; the attributable risk is 20%-2%=18%.

188. Answer is K.

189. Answer is M.

THE COCHRANE
COLLABORATION®

For example, the Cochrane logo represents seven randomised controlled trials looking at the effect of ante-partum glucocorticoid treatment to prevent respiratory distress syndrome in pre-term infants, something that was proven to be beneficial on meta analysis and is now routinely used.

The Cochrane logo is used here with permission.

190. Answer is B.

Form B can only be signed by a doctor who has attended the patient professionally within 14 days of death and has viewed the body after death; if the attending doctor is unavailable then a GP partner may sign form B.

191. Answer is A.

This form must be completed by the nearest surviving relative or executor; if this is not the case then a reason must be stated explaining this.

192. Answer is C.

Form C can only be signed by the second doctor once he has seen the body and discussed death with the first doctor, and preferably with someone who was present at the time of death (*Chief Medical Officer's Update 27*, May 2005).

Doctors filling in forms B and C should ensure that there is true independence between the two doctors completing forms A and B, that they have fully examined the body and discussed the case with each other; in addition, they should speak to a relative or other person who may have attended the deceased.

193. Answer is D.

Clindamycin is a treatment option for bacterial vaginosis, as is oral or topical metronidazole (unless pregnant in which case stick to clindamycin).

194. Answer is A.

ACE inhibitors are a good first-line choice in someone who is not black and who is less than age 55.

195. Answer is B.

Gout would be aggravated by thiazide diuretics.

196. Answer is B.

Gout would be aggravated by thiazide diuretics.

197. Answer is D.

WHO Guidelines state that beta-blockers are still indicated in the treatment of hypertension for those patients who have had an MI or who suffer from angina; they are contraindicated in asthma, COPD, heart block, diabetes and heart failure.

198. Answer is E.

Angiotensin II receptor antagonists are useful for those patients who are bothered by cough while on ACE inhibitors (*British Hypertension Society, RCP, update to NICE*, CG34 June 2006).

199. Answer is A.

This combination has been shown to produce an 83% reduction in all-cause mortality in these patients, as reported in the *BMJ* (2005; **330**: 1059–63); addition of an ACE inhibitor conferred no additional benefit despite the analysis quoted being adjusted for congestive cardiac failure.

200. Answer is B.

The 95% CI indicates that the true value has a 95% chance of lying within this range; it is represented on a Forrest Plot as a horizontal line; a longer line means a wider CI .

Questions 201-250

for answers see page 81–85

201. Cataract

Surgery for cataracts is the commonest operation in the developing world and the number one cause of blindness in the developing world; the following are all known risk factors except which one?

A   Malnutrition
B   Diabetes mellitus
C   Systemic steroids
D   UVB light
E   Gentamicin

202-207. Genetic inheritance

A   1 in 1
B   1 in 2
C   1 in 4
D   Definite
E   None of the above

Match the chance of developing the disease to each of the scenarios below.

202. Male child with both parents carriers for phenylketonuria.

203. Female child with both parents carriers for cystic fibrosis.

204. Male child of heterozygous mother carrying gene for haemophilia.

205. Female child of father with red–green colour blindness; mother is not a carrier.

206. Female child of father with red–green colour blindness; mother is a carrier.

207. Male child of parents where mother is known to have neurofibromatosis.

208. Illicit drug use during pregnancy

A 19 year old primip attends surgery for a routine antenatal test; she discloses that she regularly smokes cannabis with her boyfriend and would like to know if this is 'risky for the baby.'

Which one of the following is true?

A Smoking cannabis during pregnancy is associated with smaller birth weight
B Smoking cannabis during pregnancy is associated with gestational diabetes
C Smoking cannabis during pregnancy is associated with a decreased chance of miscarriage due to relaxed uterine tone
D Smoking cannabis during pregnancy is associated with macrosomia
E Smoking cannabis during pregnancy is associated with an increased risk of midline craniofacial defects

209-213. ENT

A Chronic secretory otitis media
B Acute otitis media with perforation
C Ramsay–Hunt syndrome
D Acoustic neuroma
E Otitis externa
F Tonsillitis

Match the diagnosis above to each of the following scenarios – each diagnosis may be used only once, or not at all.

209. A 26 year old man returns from holidaying in the Seychelles with his partner; he has a sore and itchy left ear.

210. A happy 6 year old boy, who has Down syndrome attends with bilateral conductive hearing loss.

211. A 45 year old lady presents with a right-sided lower motor neuron facial nerve palsy, imbalance and sensorineural deafness; on examination you note a crop of vesicles affecting the right ear canal

212. A 20 year old man presents with progressive sensorineural deafness, vertigo and tinnitus affecting the right ear only.

213. A 16 year old girl presents with fever, malaise, headache and bilateral ear pain which is exacerbated by swallowing.

214. Treatment of varicose veins

Which one of the following is considered the optimal treatment for varicose veins in terms of cost effectiveness?

A   Conventional saphenofemoral ligation, stripping of long saphenous vein and phlebectomies
B   Conventional sclerotherapy
C   Foam sclerotherapy
D   Phlebectomy
E   Radio-frequency and laser ablation

215. Emergency referral for varicose veins

According to NICE, which one of the following scenarios warrants an emergency referral?

A   A varicosity that has bled once and is in danger of bleeding again
B   Bleeding from a varicosity that has eroded the skin
C   An ulcer that is progressive and painful despite treatment
D   Recurrent superficial thrombophlebitis

216. Thyrotoxicosis

The following are features of thyrotoxicosis except which one?

A   Weight gain
B   Palpitations
C   Proximal myopathy
D   Increased skin pigmentation
E   Pretibial myxoedema

217. Anaemia

Causes of a microcytic anaemia include all the following except which one?

A   Iron deficiency
B   Renal failure
C   Anaemia of chronic disease
D   Thalassaemia trait
E   Pernicious anaemia
F   Sideroblastic anaemia

218. Hepatitis

Regarding hepatitis B, which one of the following statements is incorrect?
A   It is a partially double-stranded DNA virus
B   It has an incubation period of 6 days to 6 weeks
C   It can be transmitted via blood products
D   About 1% of the UK population are hepatitis B surface antigen positive
E   About 10% of infected patients become chronic carriers
F   Chronic carriage is associated with an increased risk of cirrhosis and hepatocellular carcinoma

219. Osteoporosis

Osteoporosis is associated with which one of the following?
A   Vitamin D deficiency
B   Vitamin A deficiency
C   Chronic renal failure
D   Prolonged bed rest
E   Hyperparathyroidism

220. HIV

Which one of the following statements regarding the human immunodeficiency virus is false?
A   HIV is a single-stranded RNA retrovirus
B   It induces a fall in CD4 lymphocytes, monocytes and antigen-presenting cells
C   It increases the risk of opportunistic infection
D   Patients can be infective prior to seroconversion illness at about 3 months
E   Due to the advent of modern antiretroviral treatments, the median survival with AIDS is greater than 10 years

221. Vaginal bleeding

A 22 year old lady presents with moderate left iliac fossa pain and vaginal bleeding six weeks after her last period; her pulse is 72 b.p.m. and blood pressure is 120/80; she is apyrexial; she has a past history of pelvic inflammatory disease for which she last had antibiotics 6 months ago; she has no allergies; she was seen by one of your colleagues 4 days ago when a urine test confirmed pregnancy.

Choose the next single most appropriate step in this scenario.

A   Advise bed rest and regular paracetamol until review by her own doctor tomorrow

B   Congratulate her on her pregnancy and arrange routine booking for bloods and an ultrasound scan at 12 weeks

C   Admit as a gynaecology emergency

D   Treat recurrence of PID with clarithromycin and ciprofloxacin

E   Arrange for high vaginal swabs and treat with clarithromycin and ciprofloxacin once causative agent confirmed

222. Psoriasis

Which one of the following statements regarding psoriasis is true?

A   It is aggravated by sunlight

B   It commonly spares intertriginous areas

C   Plaques have diffuse edges

D   It exhibits the phenomenon of kobnerization

E   NICE has recommended the use of tacrolismus for severe plaque psoriasis in adults who have failed to respond to standard systemic treatments

223-226. Drug interactions

A   Cranberry juice
B   Grapefruit juice
C   Marmite
D   Vodka

Match the foodstuff above with the drug it is likely to interact adversely with.

223. Disulfiram.

224. Moclobemide.

225. Simvastatin.

226. Warfarin.

227. Depression

For a 67 year old lady who had a myocardial infarction 8 months ago, and who presents with moderate depression, what would be your drug of choice?

A   Venlaflaxine
B   Nortriptyline
C   Sertraline
D   Moclobemide
E   Imipramine
F   Omega-3 fatty acid supplementation

228. Hearing tests

If an elderly lady presents with wax in her left ear causing a conductive hearing loss, you would expect which one of the following results?

A   Weber's tuning fork test lateralizing to the left and Rinne's tuning fork test being negative on the left

B   Weber's tuning fork test lateralizing to the right and Rinne's tuning fork test being negative on the right

C   Weber's tuning fork test lateralizing to the left and Rinne's being positive on the left and right

D   Weber's tuning fork test lateralizing to the right and Rinne's being positive on the left and right

229. Hearing tests

If a young man being investigated for progressive sensorineural hearing loss due to a right-sided acoustic neuroma is examined, we would expect to find which of the following results?

A   Weber's tuning fork test lateralizing to the left and Rinne's tuning fork test being negative on the left

B   Weber's tuning fork test lateralizing to the right and Rinne's tuning fork test being negative on the right

C   Weber's tuning fork test lateralizing to the left and Rinne's being positive on the left and right

D   Weber's tuning fork test lateralizing to the right and Rinne's being positive on the left and right

230. Occupational lung disease

Which one of the following statements is correct.

A   If a patient develops an occupational disease, their doctor is obliged to inform their employer in writing, with or without the patient's consent

B   If a patient develops an occupational disease, their doctor is obliged to inform RIDDOR (Reporting of Injuries, Diseases and Dangerous Occurrences Regulations) with or without the patient's consent

C   Pneumoconiosis (coal worker's lung) is not a notifiable disease

D   Irritant dermatitis (hairdresser's hands) is a notifiable disease

E   Industrial injuries disablement benefit is only payable if the effect of the injury lasts beyond the 91st day and the employee has to prove they were not to blame

231-235. HIV transmission

A   Zero risk

B   1 in 1000

C   1–3 in 100

D   1 in 300

E   5–10 in 100

Match the risk of infectivity with each exposure type described below.

231. Hugging an HIV-positive person.

232. Percutaneous exposure through needlestick injury.

233. Female having vaginal sexual intercourse with HIV-positive male.

234. Male having sexual intercourse with HIV-positive female.

235. HIV-infected urine splash in eye.

236-242. Consultation models

A Pendleton
B Roger Neighbour
C Byrne and Long
D Balint
E Eric Berne: transactional analysis
F Stott and Davis
G Helman
H Heron's interventional model
I Biomedical model

Match the description below to the single most appropriate consultation model.

236. Management of presenting complaint, management of ongoing problems, opportunistic health promotion, modification of health-seeking behaviour.

237. Connect, summarize, hand-over, safety-net, house-keeping.

238. What happened? Why? Why to me? Why now? What if I ignore it? What should I do?

239. Consultation can be instructive, informative, cathartic, confronting, supportive, catalytic.

240. History, examination, investigation, treatment.

241. This model introduced concept of heartsink patient, the doctor as a drug, the flash, the collusion of anonymity.

242. The games people play; roles of doctor and patient analysed in terms of role of parent, adult, child.

243–248. Chromosomal abnormalities

A 47XXY
B 45XO
C 5p deletion
D Trisomy 21
E Trisomy 18
F 46XX
G Trisomy 13

Match each of the phenotypes to one the genotypes.

243. Flat occiput, low set eyes with prominent epicanthic folds, single palmar crease, congenital heart disease.

244. Males, tall stature, gynaecomastia, low IQ, infertility.

245. Rocker-bottom feet, low set ears, receding chin, developmental delay, index finger overlaps 3rd digit.

246. Females, broad chests with wide-spaced nipples, increased carrying angle at elbows, lymphoedema of hands and feet.

247. Normal phenotype.

248. Microcephaly, developmental delay, alert expression, abnormal cat-like cry, moon-shaped face, marked epicanthic folds.

249. Malaria

Considering non-drug preventative interventions in adult travellers, which one of the following is likely to be most beneficial?
A Acoustic buzzers
B Air conditioning and electric fans
C Smoke
D Insecticide treated clothing and/or nets
E Aerosol insecticides

250. Opioid detoxification

Considering pharmacological interventions in opioid detoxification, which one of the following is true?
A Clonidine or dihydrocodeine may be used routinely
B Buprenorphine is a second-line treatment
C Methadone is a second-line treatment
D Detoxification should normally last up to 12 months in the community
E Lofexidine may be considered for people with mild dependence

Answers 201-250

201. Answer is E.

Gentamicin is ototoxic.

The main extrinsic factors associated with cataract formation in the developed world are smoking, diabetes and use of systemic steroids. Additional factors in the developing world include malnutrition, acute dehydrating disease and cumulative exposure to UVB sunlight (*BMJ*, 2006; **333**: 128–32).

202. Answer is C.

Phenylketonuria is inherited in an autosomal recessive manner.

203. Answer is C.

Cystic fibrosis is inherited in an autosomal recessive manner (as are sickle cell disease and glycogen storage diseases).

204. Answer is B.

Haemophilia is inherited as a sex-linked disorder.

205. Answer is E.

Red–green colour blindness is inherited as a sex-linked disorder, as is fragile-X and Duchenne muscular dystrophy.

206. Answer is B.

207. Answer is B.

Neurofibromatosis, myotonic dystrophy, and Marfan syndrome are all autosomal dominant.

208. Answer is A.

Smoking cannabis during pregnancy is associated with low birth weight, smaller head circumference, greater risk of miscarriage, especially in those who use marijuana regularly (more than six times per week); these babies, when born, may undergo withdrawal-like symptoms.

Marijuana can reduce fertility in both men and women, making it difficult to conceive in the first place.

However, many women who take illicit drugs also take alcohol and tobacco, and may engage in other unhealthy behaviours, putting their pregnancy at risk; therefore it is difficult to be certain what can be attributed solely to the illicit substances.

209. Answer is E.

210. Answer is A.

211. Answer is C

212. Answer is D.

213. Answer is F.

214. Answer is A.
Conventional varicose vein surgery is a clinically and cost-effective treatment; laser and radio-frequency treatment replace traditional stripping and most varicosities still need to be treated by sclerotherapy or phlebectomy (*BMJ*, 2006; **333**: 287–92).

215. Answer is B.
Bleeding from a varicosity that has eroded the skin should be referred as an emergency; one which has bled once and is in danger of doing so again should be referred urgently; a progressive, painful ulcer not responding to treatment should be referred quickly, and recurrent superficial thrombophlebitis should be referred as a routine out-patient appointment (*BMJ*, 2006; **333**: 287–92).

216. Answer is A.
Weight gain is a feature of an underactive thyroid.

217. Answer E is false.
Pernicious anaemia leads to a macrocytic anaemia due to vitamin B12 deficiency.

218. Answer B is false.
The correct incubation period for hepatitis B is 2–6 months.

219. Answer is D.
Osteoporosis is the result of bone loss, caused by a change in factors that regulate bone cell metabolism; one of these is prolonged bed rest. Bone loss generally occurs at rate of 1–2% per month during prolonged periods of bed rest; physical activity can delay the progression of osteoporosis by slowing the rate of bone mineral density reduction.

Osteomalacia is a defect in the process of mineralization of bone, nearly always due to vitamin D deficiency.

220. Answer E is false.
The virus increases the risk of opportunistic infection (e.g. pneumocystis carinii) and malignancy (e.g. Kaposi's sarcoma). Patients may have asymptomatic or symptomatic disease for several years before developing AIDS. The median survival with full AIDS is less than 2 years.

221. Answer is C.

This girl may well have an ectopic pregnancy – a gynaecological emergency which needs emergency admission.

222. Answer D is true.

Other skin disorders known to kobnerize – that is, the appearance of skin lesions on previously normal skin after an incident of skin injury or trauma include: lichen sclerosis, lichen planus and erythema multiforme.

NICE (July 2006) recommended use of etanercept (a cytokine inhibitor) for the use of severe plaque psoriasis in patients who have not responded to standard systemic treatments.

Tacrolismus is licensed for use in moderate to severe eczema.

223. Answer is D.

Disulfiram is used as an adjunct to treatment of alcohol dependence and gives rise to extremely unpleasant systemic reactions after ingestion of even a small amount of alcohol.

224. Answer is C.

Patients taking monoamine oxidase inhibitors such a moclobemide should not take substances such as Marmite, Bovril, mature cheese, or pickled herring, because of the risk of a dangerous hypertensive crisis.

225. Answer is B.

Grapefruit can increase the risk of rhabdomyolysis as a side effect of taking simvastatin.

226. Answer is A.

Cranberry juice can raise the INR of patients on warfarin.

227. Answer is C.

When initiating anti-depressant treatment in patients with recent myocardial infarction or unstable angina, sertraline is the treatment of choice and has the best evidence base (*NICE CG23*, Dec 2004, April 2007).

There is evidence to show that taking long chain omega-3 fatty acids may help to relieve depression when given in addition to anti-depressant therapy, but the evidence is not strong enough to recommend routine supplementation.

228. Answer is A.

229. Answer is C.

These hearing tests use a 512 Hz tuning fork and compare air conduction (AC) and bone conduction (BC).

In Rinne's test, the struck tuning fork is held in front of the ear (AC) and on the mastoid process (BC) to see which the patient can hear for the longest; usually the patient is AC>BC and this is termed Rinne's +ve. If there is some impediment to AC, e.g. conductive hearing loss as in glue ear, then BC>AC as sound waves will be conducted through bone to the normally functioning sensorineural system. This is termed Rinne's –ve. A false positive occurs in sensorineural hearing loss when BC appears >AC but only because sound waves are transferred to the opposite ear through the skull bone. The chances of a false positive can be eliminated using a masking technique on the ear that is not being tested.

In Weber's test a struck tuning fork is held in the middle of the forehead; sound should conduct equally to both ears in a patient with normal hearing; in conductive deafness, e.g. of the right ear, sound will localize to the right ear; however, if the right ear has sensorineural deafness, then sound will not be picked up there but will be heard loudest in the left, normally functioning ear.

230. Answer is D.

Patient must give consent to inform employer; alternatively, patient may give consent to inform RIDDOR instead; 'industrial' covers all forms of work and is payable even if employee is partially or wholly to blame; effect of injury must persist beyond 91 days to qualify (*Oxford Handbook General Practice*, 2005; see also www.riddor.gov.uk).

231. Answer is A.

232. Answer is D.

233. Answer is E.

234. Answer is C.

235. Answer is B.

236. Answer is F.

237. Answer is B.

238. Answer is G.

239. Answer is H.

240. Answer is I.

241. Answer is D

242. Answer is E

243. Answer is D.

Trisomy 21 is also known as Down syndrome.

244. Answer is A.

47XXY is Kleinfelter's syndrome.

245. Answer is E.

Trisomy 18 is Edward's syndrome.

246. Answer is B.

45XO is Turner's syndrome.

247. Answer is F.

Normal genotype for males is 46XY and for females 46XX.

248. Answer is C.

5p deletion (deletion of end of short arm of chromosome 5) results in cri-du-chat syndrome.

249. Answer is D.

Insecticide-treated clothing and insecticide-treated nets are likely to be beneficial; the effectiveness of the others is unknown. There is a consensus that topical insect repellents containing DEET reduce the risk of insect bites, although few studies have been done (*BMJ Clinical Evidence: Malaria – Prevention in Travellers*, October 2006).

250. Answer is E.

One should offer either methadone or buprenorphine as first-line treatments, having taken the preference of the service user into consideration; clonidine or dihydrocodeine should not be used routinely. Lofexidine may be considered for people with mild or uncertain dependence who have made an informed and clinically appropriate decision. Detoxification in the community should normally last up to 12 weeks (*NICE CG51 and CG52*, July 2007).

Algorithm Questions 1-53

for answers see page 95–103

1-8. Medical management of osteoporosis

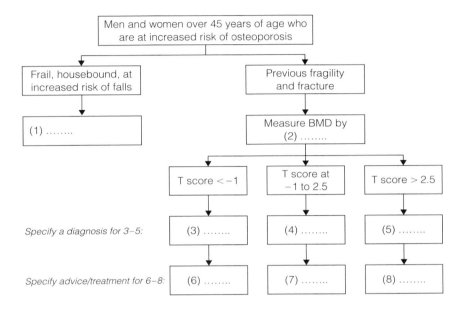

For each of the numbered gaps above, select one option from the list below to complete the algorithm, based on current evidence.

A   Calcium plus vit D supplements; falls assessment and advice; hip protectors

B   Dual Energy X-ray Absorptionometry

C   CT scan

D   MRI scan

E   Normal

F   Osteopenia

G   Osteoporosis

H   Osteomalacia

I   Rickets

J   Reassure; lifestyle advice (nutrition, vitamin D, Ca); regular weight-bearing exercise; reduce smoking; reduce alcohol

K   Lifestyle advice and treat if previous fracture

L   Lifestyle advice and offer treatment

9-12. Medical management of actinic keratoses

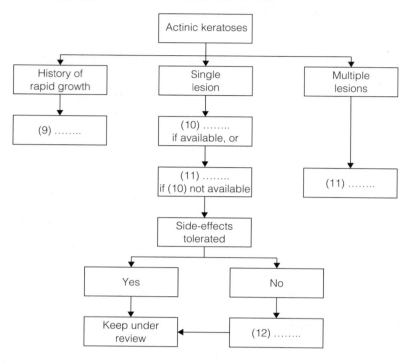

For each of the numbered gaps above, select one option from the list below to complete the algorithm, based on current evidence.

A 3% diclofenac cream

B 10% diclofenac cream

C Cryotherapy

D Urgent referral

E 5% fluorouracil cream

13-17. Medical management of a patient with atrial fibrillation

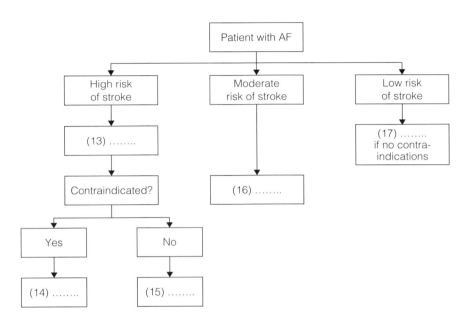

For each of the numbered gaps above, select one option from the list below to complete the algorithm based on current evidence. Each option may be used once, more than once, or not at all.

A   Consider anticoagulation with warfarin

B   Regular low dose of subcutaneous heparin

C   TED stockings

D   INR target: 2.5 (range 2.0–3.0)

E   INR target: 1.5 (range 1.0–2.0)

F   INR target: 3.5 (range 3.0–4.0)

G   Aspirin 75–300 mg o.d. if no contraindications

H   Consider aspirin or warfarin on a case by case basis

18-24. Medical management of a breast lump

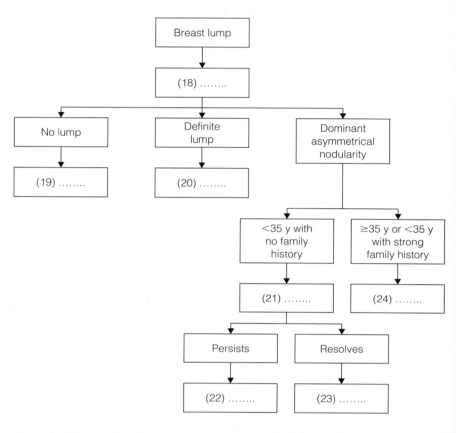

For each of the numbered gaps above, select one option from the list below to complete the algorithm, based on current evidence.

A   History and examination
B   Refer
C   Reassure
D   Review at 6 weeks
E   Biopsy
F   Wide local excision
G   Tamoxifen

25-31. Symptoms and signs suggestive of chronic heart failure

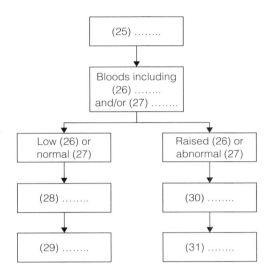

For each of the numbered gaps above, select one option from the list below to complete the algorithm, based on current evidence.

A History and clinical examination
B Brain natiuretic peptide
C Atrial natiuretic peptide
D Chest X-ray
E ECG
F CT scan
G CHF excluded
H CHF possible
I Consider other causes for symptoms
J Refer to echo to assess cardiac function further

32-37. Pharmacological management of facial hirsutism

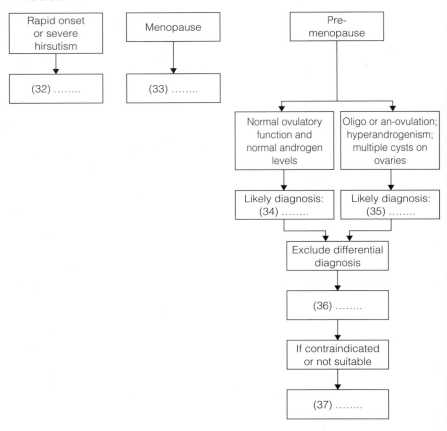

For each of the numbered gaps above, select one option from the list below to complete the algorithm, based on current evidence.

A Cyproterone acetate / ethinyl estradiol
B Refer
C PCOS
D Idiopathic hirsutism
E Eflornithine cream
F Thiazide diuretics

38-45. Choosing drugs for patients newly diagnosed with hypertension

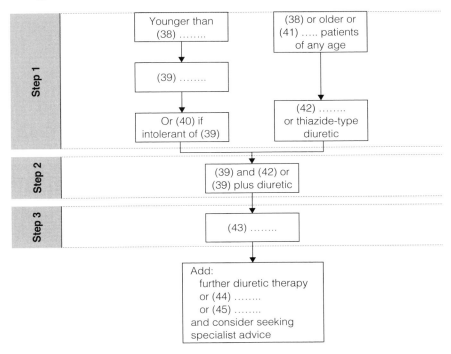

For each of the numbered gaps above, select one option from the list below to complete the algorithm, based on current evidence.

A   45 years of age
B   50 years of age
C   55 years of age
D   ACE inhibitor
E   Angiotensin II receptor antagonist
F   Black
G   Eurasian
H   Asian
I   Chinese
J   Ca antagonist
K   ACE inhibitor plus Ca antagonist plus thiazide diuretic
L   β blocker
M   ACE inhibitor plus β blocker plus thiazide diuretic
N   α blocker

46-53. Management of an unconscious adult

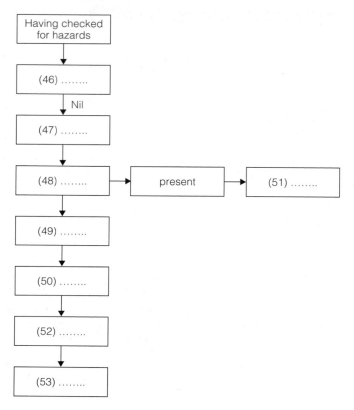

For each of the numbered gaps above, select one option from the list below to complete the algorithm, based on current evidence.

A Assess for 10 sec
B Check responsiveness
C Two effective breaths
D Open airway by head tilt, chin lift
E Check breathing
F Assess for 1 min
G If breathing, place in recovery position
H Continue rescue breathing, check circulation every minute
I Begin chest compressions
J CPR rate: 5 chest compressions per 1 rescue breath
K CPR rate: 15 : 2
L Call 999 for help
M 30 chest compressions
N CPR rate: 30 : 2

Answers to algorithm questions 1–53

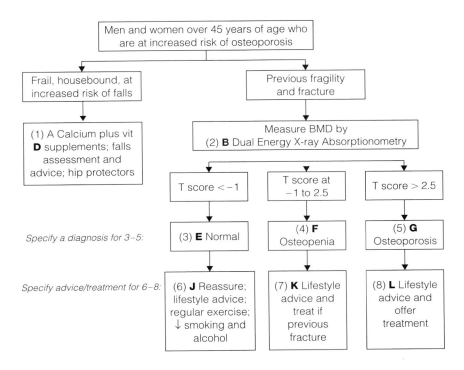

1. **Answer is A.**
2. **Answer is B.**
3. **Answer is E.**
4. **Answer is F.**
5. **Answer is G.**
6. **Answer is J.**
7. **Answer is K.**
8. **Answer is L.**

Answers to questions 1–8 are all drawn from Royal College of Physicians and Bone and Tooth Society of Great Britain (July 2005): *Osteoporosis – clinical guidelines for prevention and treatment.*

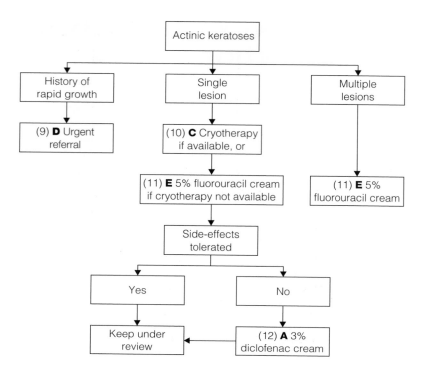

9. **Answer is D.**
10. **Answer is C.**
11. **Answer is E.**
12. **Answer is A.**

Answers to questions 9–12 drawn from Working Party Guidelines (2004): *The primary and shared care management of actinic keratoses.*

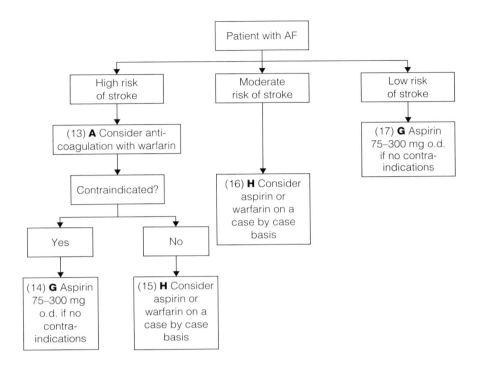

13. **Answer is A.**
14. **Answer is G.**
15. **Answer is H.**
16. **Answer is H.**
17. **Answer is G.**

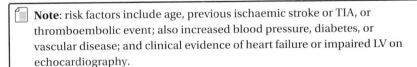
Note: risk factors include age, previous ischaemic stroke or TIA, or thromboembolic event; also increased blood pressure, diabetes, or vascular disease; and clinical evidence of heart failure or impaired LV on echocardiography.

Answers to questions 13–17 are drawn from *NICE Guidelines* (June 2006)

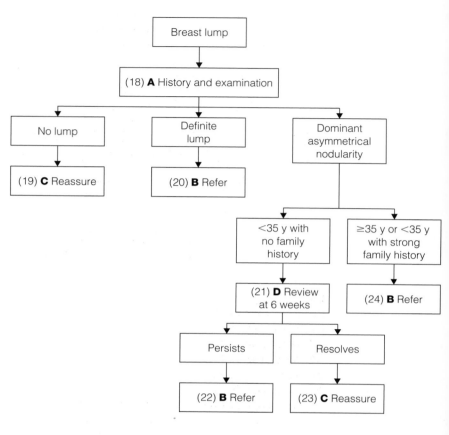

18. **Answer is A.**
19. **Answer is C.**
20. **Answer is B.**
21. **Answer is D.**
22. **Answer is B.**
23. **Answer is C.**
24. **Answer is B.**

Answers to questions 18–24 are all drawn from:

- NHS Cancer Screening Programmes and Cancer Research UK (2003) *Guidelines for referral of patients with breast problems*
- *NICE Clinical Guidelines* (August 2002) *Improving outcomes in breast cancer*

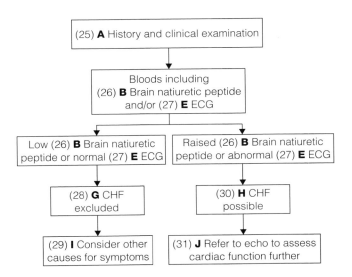

25. Answer is A.
26. Answer is B.
27. Answer is E.
28. Answer is G.
29. Answer is I.
30. Answer is H.
31. Answer is J.

Answers to questions 25–31 are all drawn from:

- *NICE Clinical Guidelines* (July 2003)
- *European Heart Journal* 2005; **26**: 1115–40 – *Guidelines for the diagnosis and treatment of chronic heart failure* – European Society of Cardiology

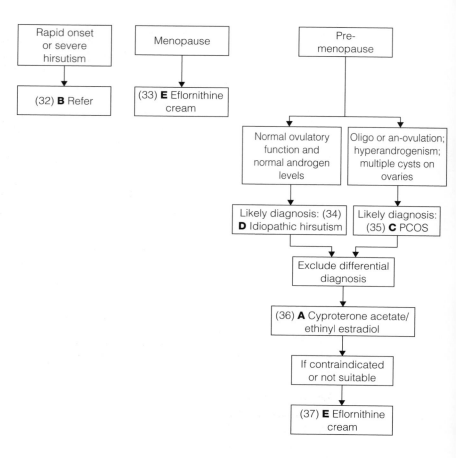

32. **Answer is B.**
33. **Answer is E.**
34. **Answer is D.**
35. **Answer is C.**
36. **Answer is A.**
37. **Answer is E.**

All answers drawn from www.eGuidelines (2007) *Medical management of facial hirsutism: working party guidelines.*

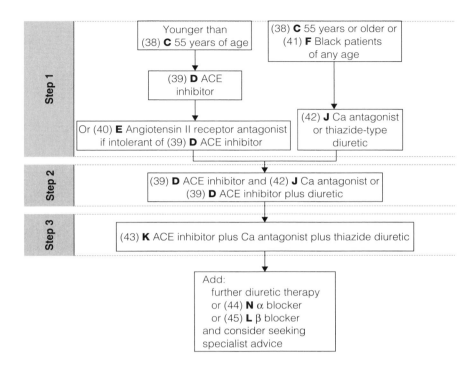

Step 1

Younger than
(38) **C** 55 years of age

(38) **C** 55 years or older or
(41) **F** Black patients
of any age

(39) **D** ACE
inhibitor

Or (40) **E** Angiotensin II receptor antagonist
if intolerant of (39) **D** ACE inhibitor

(42) **J** Ca antagonist
or thiazide-type
diuretic

Step 2

(39) **D** ACE inhibitor and (42) **J** Ca antagonist or
(39) **D** ACE inhibitor plus diuretic

Step 3

(43) **K** ACE inhibitor plus Ca antagonist plus thiazide diuretic

Add:
further diuretic therapy
or (44) **N** α blocker
or (45) **L** β blocker
and consider seeking
specialist advice

38. Answer is C.
39. Answer is D.
40. Answer is E.
41. Answer is F.
42. Answer is J.
43. Answer is K.
44. Answer is N.
45. Answer is L.

Answers to questions 38–45 are all drawn from *NICE Guidelines* (June 2006).

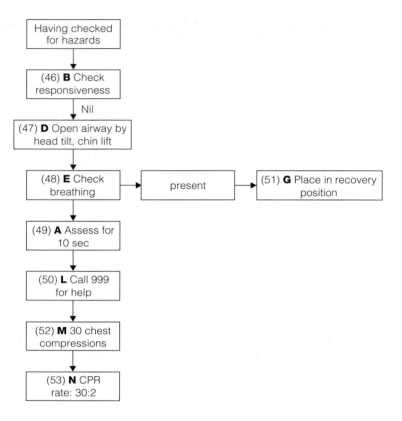

46. Answer is B.
47. Answer is D.
48. Answer is E.
49. Answer is A.
50. Answer is L.
51. Answer is G.
52. Answer is M.
53. Answer is N.

Answers to questions 46–53 are all drawn from *Resuscitation Council (UK) Guidelines* (2005).

Picture Questions 1-29

for answers see page 117–121

1. Skin rash

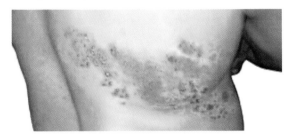

With regard to this rash choose one correct answer from the list below:

A typically painless

B typically seen in fit healthy young men

C it is due to herpes simplex virus type 2

D topical lidocaine is not recommended as a first line agent in the treatment of post rash complications

E reduced in terms of duration and severity of pain if systemic antiviral treatment is started within 4 to 8 days after presentation

2. Skin rash

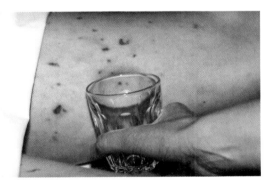

Which one of the following statements is true?

A Early signs of this disease in children include leg pains, cold hands/feet and a mottled skin colour

B Oral penicillin should be given before hospital admission by the GP

C This non-blanching haemorrhagic purple rash is due to chicken pox virus

D In children, the classic symptoms of this disease (i.e. rash, headache and impaired consciousness) always appear within the first 2 hours of the illness

E IV penicillin should be given before hospital admission by the GP

F The GP should then personally arrange contact tracing and prophylaxis for the family and kissing contacts.

3-6. Systematic review and meta analysis

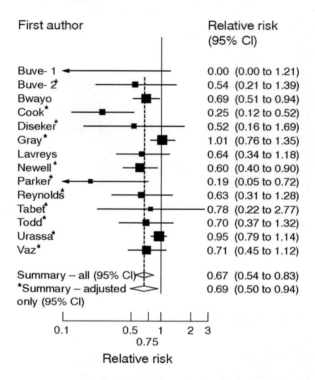

First author | **Relative risk (95% CI)**

First author	Relative risk (95% CI)
Buve- 1	0.00 (0.00 to 1.21)
Buve- 2	0.54 (0.21 to 1.39)
Bwayo	0.69 (0.51 to 0.94)
Cook*	0.25 (0.12 to 0.52)
Diseker*	0.52 (0.16 to 1.69)
Gray*	1.01 (0.76 to 1.35)
Lavreys	0.64 (0.34 to 1.18)
Newell*	0.60 (0.40 to 0.90)
Parker*	0.19 (0.05 to 0.72)
Reynolds*	0.63 (0.31 to 1.28)
Tabet*	0.78 (0.22 to 2.77)
Todd*	0.70 (0.37 to 1.32)
Urassa*	0.95 (0.79 to 1.14)
Vaz*	0.71 (0.45 to 1.12)
Summary – all (95% CI)	0.67 (0.54 to 0.83)
*Summary – adjusted only (95% CI)	0.69 (0.50 to 0.94)

Relative risk (x-axis: 0.1, 0.5, 0.75, 1, 2, 3)

A Horizontal line (there are 14 shown)
B Square in middle of each horizontal line
C Width of each horizontal line
D Solid vertical line down middle
E Diamond below all horizontal lines (there are two here)

Considering the systematic review and meta analysis data shown above, match the definitions to the description given; each definition may be used only once, and not all of them are used.

3. This sign represents pooled data from all trials shown.

4. This represents the 95% confidence interval of this estimate.

5. This is the line of no effect and is associated with a relative risk of 1.0.

6. This corresponds to each trial and shows the relative risk of the condition as a result of the intervention

7. Device failure

Which one of the following does not affect the failure rate of this device.

A   Petroleum jelly
B   Vaseline
C   Baby oil
D   Clotrimazole cream 1%
E   KY jelly

8. Ophthalmology

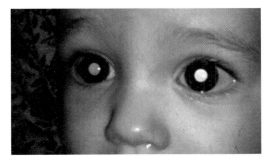

The mother of a two year old child brings her to see you; she has brought with her a number of recent photos, including the one shown above; the child is well and asymptomatic; on examination with an ophthalmoscope you are unable to elicit a red reflex in the left eye.

Which one of the following is true?

A   The child should be reviewed by yourself in two weeks
B   It most likely be a problem with the camera which should be replaced and the child re-photographed
C   The child should be referred routinely
D   The condition could be lethal
E   The condition is never hereditary

9. Ophthalmology

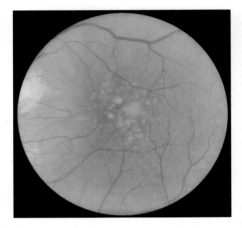

Which one of the following statements concerning this photograph is correct?

A   This is a photo showing panretinal photocoagulation scars in a patient with diabetic retinopathy

B   The white spots are small haemorrhages

C   This picture is typical of wet age-related macular degeneration

D   This picture is typical of dry age-related macular degeneration

E   The deposits are covering the optic nerve

10. ECG

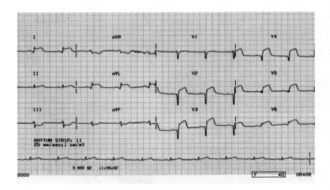

A 40 year old man collapses with chest pain; his ECG shown above displays which one of the following.

A   Acute anterior myocardial infarction

B   Acute postero–inferior myocardial infarction

C   Old anterior myocardial infarction

D   Pulmonary embolism

E   Normal ECG

11. Facial Weakness

A fit and healthy, non-smoking 26 year old man attends with sudden onset of weakness affecting the left side of his face; he has a smooth forehead and is dribbling from the left side of his mouth; he is unable to close his left eye. His grandfather recently had a stroke at the age of 66 years. He looks as shown in the diagram above.

Further examination reveals a crop of vesicles affecting the external auditory meatus of the left ear and he has developed a sensitivity to loud sounds.

Which one of the following is true?
A   This man has had a stroke and needs to be referred to a stroke unit
B   This man has herpes zoster oticus
C   He has an upper motor lesion affecting the facial cranial nerve
D   He has a lower motor lesion affecting the tenth
E   Taste disturbance affecting the ipsilateral 2/3 of the tongue is uncommon

12. Oral mucosal discolouration

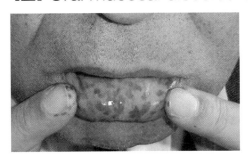

What does this photograph illustrate?
A   Peutz Jeghers syndrome
B   Acanthosis nigricans
C   Hereditary haemorrhagic telangiectasia
D   Addison's disease

13. Skin rash

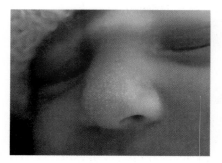

A first time mum attends with her two week old child worried about a rash (as shown in the photograph above) on her baby's nose.

Which one of the following options is the correct management of these small white spots?

A   Mild topical steroid for two weeks
B   1% clotrimazole
C   Pierce with orange stick
D   Nothing
E   Curettage

14. Musculoskeletal

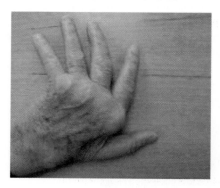

Considering the photograph above, which one of the following statements is true?

A   The patient is unlikely to have systemic involvement
B   Early diagnosis and treatment are crucial to avoid irreversible damage to the joints
C   Anti-cyclic citrullinated peptide antibodies are not highly specific for this disease
D   If a patient is treated with biological agents, they need to be informed that treatment will be lifelong and they are unlikely to be taken off such medication once it has been initiated
E   The patient is more likely to be male than female.

15. Dermatology

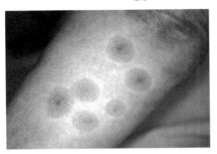

An 18 year old man who was prescribed a course of antibiotics for a sore throat last week, saw one of your colleagues yesterday when he developed these lesions on his forearms; they appeared suddenly, starting as small flat red spots that enlarged over the next day or so, the central area cleared and now looks pale purple; they are minimally itchy and not painful. However, the rash has spread proximally and although he is well, he is concerned that he may have chicken pox.

Which one of the following best describes this picture?

A   Erythema nodosum
B   Erythema multiforme
C   Vitiligo
D   Erythema ab igne
E   Erythrasma

16. Dermatology

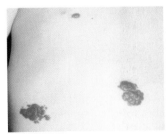

An anxious mother brings her 4 week old daughter; she has developed these small lesions on her abdomen; the mother is certain they were not there when the baby was born and is worried they are getting bigger. Physical examination of the baby is otherwise normal.

Which one of the following should you advise in terms of treatment?

A   Topical fusidic acid
B   Salicylic acid
C   Clotrimazole
D   Topical silver nitrate
E   Nothing

17. ENT

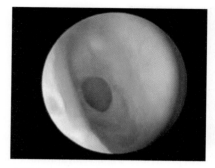

Which one of the following best describes the photograph above?

A Perforated ear drum

B Normal tympanic membrane

C Cholesteatoma

D Grommet *in situ*

E Wax

18. Ophthalmology

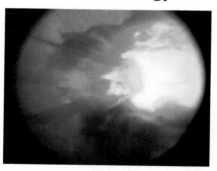

A 61 year old lady, who has been a poorly controlled diabetic for 26 years, comes to you worried; she awoke that morning with a sudden onset of misty vision in her right eye: she describes waking up and her vision being like looking through a car windscreen on a rainy day; there was no pain; on examination she is able to see hand movements only with the right eye and VA in the left is 6/9. On ophthalmoscopy, you see the image as shown; the left fundus reveals proliferative retinopathy. BP is 160/90.

Which one of the following best describes what has happened?

A Vitreous haemorrhage

B Acute cataract

C Acute glaucoma

D Central retinal artery thrombosis

E Hysterical blindness

19. Toe nails

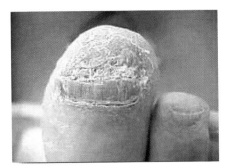

Which one of the following is the best systemic treatment for this condition?

A   Terbinafine
B   Flucloxacillin
C   Fluconazole
D   Tea tree oil
E   Griseofulvin
F   Itraconazole

20. Dermatology

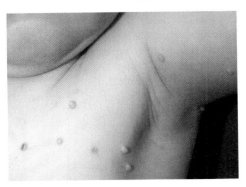

A mother brings her otherwise well 3 year old to see you; she has developed these spots on her chest and, although they do not bother the child, the mother is worried because initially there were only three but they have increased in number. The school nursery has asked that the child be excluded and have told the mother that the little girl has a highly contagious virus; both mother and daughter swim regularly at the local pool.

Which one of the following best describes the condition?

A   Seborrhoeic warts
B   Molluscum contagiosum
C   Chicken pox
D   Smallpox

21. Rheumatology

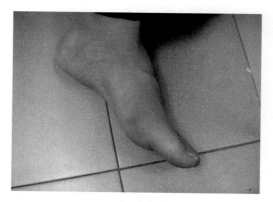

A 58 year old man has requested a house call because of sudden onset intense pain affecting the big toe; he describes it as an excruciating, gnawing sensation and is unable to tolerate even the bed covers on it; he is otherwise fit and well apart from "a touch of blood pressure" for which he takes bendroflumethiazide 2.5 mg o.d. The clinical picture is as shown in the photograph above.

Which one of the following statements is correct?

A   In the acute phase, where there are no contraindications, fast-acting i.m. NSAIDs are the drugs of choice

B   Colchicine is effective and works as quickly as NSAIDs

C   Allopurinol may be commenced during the acute attack

D   In overweight patients, dieting should be encouraged; a high protein, low carbohydrate diet (e.g. Atkins) is ideal

E   Overall protein intake should be restricted

22. ENT

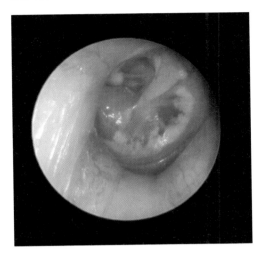

An 18 year old boy attends complaining of deafness, affecting the right ear; this has been getting worse over the past few months; he gives a long history of recurrent otitis media affecting both ears, worse on the right, for which he had grommets on two occasions as a child; there is no tinnitus, vertigo, nystagmus or systemic upset. Examination of the left drum is normal, the right drum is as shown on the photograph above.

Webers test localizes to the right ear and Rinnes test is louder behind the right ear than in front, and louder in front of the left ear than behind.

Which one of the following is true?

A He is likely to have tympanosclerosis

B He should have the deposits on the typanic membrane scraped away with a sharp implement

C He has right sensorineural deafness

D He has left conductive hearing loss

23. ECG

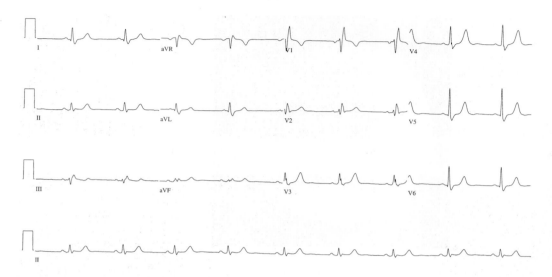

Which one of the following is false in this ECG of a 22 year old male undertaken during a routine medical.

A   The ECG shows left bundle branch block

B   The abnormality shown could be a normal variant in some people

C   The patient is in sinus rhythm

D   The patient needs routine referral to cardiac outpatients for a pacemaker

24-29: Dermatology

24.

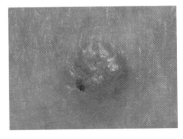

25.

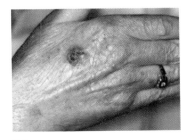

26.

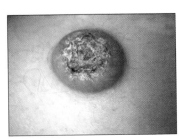

27.

28.

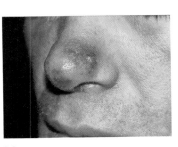

29.

Match the skin lesions below to the correct picture:

A   Basal cell carcinoma

B   Seborrhoeic keratoses

C   Malignant melanoma

D   Keratoacanthoma

E   Squamous cell carcinoma

F   Kaposi's sarcoma

Picture question answers 1-29

1. Answer D is true.

Shingles is due to re-activation of chicken pox virus; the rash is preceded by pain and the affected area is usually hyperaesthetic; the pain can be severe. It can occur at any age, but more commonly affects the elderly and immunocompromised.

Oral antivirals such as acyclovir are only effective if started within 48–72 hours of onset (*BNF* Sept 2007).

There is insufficient evidence to recommend topical lidocaine as a first-line agent in the treatment of post-herpetic neuralgia; it may benefit some patients but there is stronger evidence for use of other classes of drugs, e.g. gabapentin (Khaliq, Alam and Puri. Topical lidocaine for treatment of post herpetic neuralgia. *Cochrane Database of Systematic Reviews*, Issue 2).

2. Answer A is true.

An important study in the *Lancet* (2006; **367**: 397–403) showed that the classic symptoms of meningococcal disease presented late (median onset 13–22 hours); however, at a median time of 8 hours, 72% had developed early signs of sepsis (leg pains, etc.).

Penicillin i.m. should be given before urgent transfer to hospital (*BNF* Sept 2007).

Contact tracing and prophylaxis is undertaken by the local public health department.

3. Answer is E.

As the diamond does not cross the line of no effect, this meta analysis would indicate that circumcision may be beneficial in reducing the risk of syphilis, chancroid and genital herpes.

4. Answer is C.

If the confidence interval of the result, i.e. the horizontal line, crosses the vertical line of no effect, that can mean either there is no difference in outcome as a result of the intervention, or, that the sample size was too small for us to be confident of where the true result lies.

5. Answer is D.

The line of no effect as shown here is associated with a relative risk of 1.0.

6. Answer is A.

Each trial is represented by a line.

The area of the square is proportional to the statistical strength of the evidence in the study. The more subjects in the study, the larger the square.

7. Answer is E.

CSM advice in *BNF* (Sept 2007) is that oil and oil-based preparations such as vaseline and petroleum jelly are likely to cause damage to condoms and contraceptive devices made of rubber, making them less effective as a barrier method of contraception and as a protection from STDs including HIV; KY jelly is a non-spermicidal water-based lubricating agent that does not affect the latex; women should be advised to use spermicidal cream/pessaries with their diaphragms.

The *BNF* also states that clotrimazole cream can affect latex condoms/diaphragms.

8. Answer is D.

The child is presenting with leucocoria (white pupillary reflex); childhood leucocoria must be referred urgently as it may be due to retinoblastoma, a life-threatening tumour of early childhood. Prognosis is dependent on early detection and can range from cure to death. It may be hereditary or sporadic.

9. Answer is D.

The photo shows a picture typical of dry age-related macular degeneration with drusen affecting the macula.

10. Answer is A.

ST elevation in the anterior leads indicates that this man is having an acute anterior MI.

11. Answer B is true

Bell's palsy is a lower motor lesion affecting the facial (seventh) cranial nerve; the whole of the side of the face is affected; the frontalis muscle is not spared as it is in an upper motor neurone lesion, e.g. CVA. The facial nerve supplies taste fibres to the anterior 2/3 of the tongue via the chorda tympani.

In herpes zoster oticus (Ramsay Hunt syndrome) severe pain in the ear precedes the facial nerve palsy; zoster vesicles appear in the external ear canal and on the soft palate.

12. Answer is A.

Peutz Jeghers syndrome is characterised by mucocutaneous dark freckles on lips, oral mucosa, face, palms and soles (look closely at the hands!); it is autosomal dominant; there are benign intestinal polyps as part of the syndrome which can bleed or cause obstruction; malignancy occurs in 3%.

13. Answer is D.

These are milia; small white raised spots that are commonly seen in the neonate; reassure the mother that they will resolve on their own and that no treatment is required.

14. Answer B is true.

This patient is showing classical signs of rheumatoid arthritis with swelling of the fingers, metacarpal joints and ulnar deviation; RA is a chronic disease with a female : male ratio of 2 : 1 which affects not just the synovium joint but numerous other organs (pericarditis, pleurisy, anaemia, vasculitis, eye involvement, Felty's syndrome, etc.).

Research has shown that early diagnosis and treatment can improve disease outcome and early data have shown that if inflammation can be suppressed (by disease-modifying anti-rheumatic drugs, or, even more efficiently, by biological agents) at onset of disease, therapy can be withdrawn; the patient is in 'remission'. Many autoimmune diseases are positive for rheumatoid factor; it can also be found in the blood of 'healthy' asymptomatic individuals; it is sensitive but not specific for RA; anti-CCP antibodies are more specific (95% compared to 85%) (*BMJ* 2006; **332**:152-155; *Ann Intern Med* 2007; **146**:797-808).

15. Answer is B.

The picture is of erythema muliforme with its typical target or iris lesion; it may be related to his previous (presumed) strep throat or to penicillin, if he was prescribed this; it is usually self limiting and requires only symptomatic treatment; a severe form, Stevens Johnsons syndrome is more serious.

- Erythema nodosum: painful, raised red lesions on shin fronts.
- Vitilgo: white patches.
- Erythema ab igne: chronic inflammation, hyperpigmentation due to repeated exposure to external heat source, e.g. shins of elderly patients, as a result of sitting too near to fire.
- Erythrasma: infection by *Corynebacterium minutissimum*, intertriginious areas (groin, axilla) – brown discolouration may be treated with erythromycin as the drug of choice; but can use other antibacterial and / or antifungal agents.

16. Answer is E.

This baby has a strawberry naevus, also known as a capillary haemangioma; they are very common and the mother should be reassured that although the mark will get larger as the baby grows, it will start to regress of its own accord after the age of 3 or 4.

17. Answer is A.

Can be due to infection, sudden loud noise, barotrauma, insertion of sharp objects into ear; usually heals on its own; may need ENT referral if not healing, or if perforation is large, for tympanoplasty; advise patient to avoid getting water in – e.g., no swimming or diving; protect ear with cotton wool/ear plugs while showering/bathing, until healed.

18. Answer is A.

In diabetics, the most common cause of a vitreous haemorrhage is leakage from new vessels in proliferative retinopathy; central retinal artey thrombosis also presents as acute, painless partial visual loss, often in diabetics and hypertensives, but the retina is extremely pale, whitened by the ischaemia, except for a cherry red spot at the fovea where the retina is much thinner and the choroid shows through; acute glaucoma is painful. In hysterical blindness the fundus appears normal.

19. Answer is A.

The *BNF* (Sept 2007) states that although both terbinafine and (pulsed courses of) itraconazole have replaced griseofulvin for the treatment of onychomycosis, terbinafine is considered the drug of choice; mild localized infections may respond to topical therapy.

20. Answer is B.

Molluscum contagiosum is caused by a wart virus and, although it is very infectious, the government's HPA (Health Protection Agency) specifically advises that it 'isn't a serious condition and probably not highly contagious in schools .... no exclusion from school, work or swimming pools is necessary ..... although common sense measures such as avoiding sharing towels .... may reduce transmission'.

21. Answer is E.

This man has gout; oral NSAIDs are recommended for first-line treatment (Jordan *et al. Rheumatology*, 2007; British Society for Rheumatology and British Health Professionals in Rheumatology Guideline for the Management of Gout); colchicine is an effective alternative but is slower to work than NSAIDs; allopurinal should not be commenced during the acute attack.

There is evidence that obesity is linked with gout (e.g. Sutaria *et al. Rheumatology*, 2006; **45**: 1422–31); however, weight loss should be gradual and starvation or high protein diets which can elevate urate levels should be avoided. Thiazide diuretics aggravate gout and a medication review to change this to another anti-hypertensive should be considered.

22. Answer is A.

White patches on the tympanic membrane in someone with a long history of local infection (and/or trauma, e.g. grommets), are likely to be calcium deposition; if severe they can cause a conductive hearing loss; these deposits should not be removed as there is danger of perforating the ear drum.

23. Answer is A.

The patient has right bundle branch block, which although it is found in heart disease, can be a normal variant in otherwise healthy patients.

24. Answer is B.

Seborrhoeic warts are seen frequently in the elderly, especially on the chest; flat-topped, stuck-on appearance; benign; no treatment necessary unless bothersome.

25. Answer is A.

Basal cell carcinoma is the commonest malignant skin tumour; classically it has a pearly nodule with a rolled edge and surface telagiectasia; local invasion can be very destructive but metastases are rare. Occurrence predominantly on face and other sun-exposed areas.

26. Answer is E.

Squamous cell carcinoma – locally invasive and metastasize to local lymph nodes; can present as keratotic lump, a rapidly growing polypoid mass or, as seen here, a cutaneous ulcer; again, sun-exposed sites, especially lips. Also related to pipe and cigarette smoking.

27. Answer is C.

Malignant melanoma is the most dangerous of malignant skin tumours; occur in younger patients, incidence increasing; malignant melanoma can arise in pre-existing melanocytic naevi.

28. Answer is D.

Keratoacanthoma, rapidly growing lesion – enlarges over 6–8 weeks; round tumour with rolled edges and a prominent keratin plug; ultimately shrinks away leaving a small puckered scar.

Keratoacanthoma is also regarded in the dermatology community as a squamous cell carcinoma as it can behave in exactly the same way and the histology can be almost identical. For the purposes of management, keratoacanthoma should be managed as a squamous cell carcinoma.

29. Answer is F.

Kaposi's tumour, purplish plaques and nodules; classically in Ashkenazi Jews and northern Italians; more aggressive form seen in AIDS.

Appendix

Photograph permissions

Picture question 1
Reproduced from the National Library of Medicine.

Picture question 2
Reproduced courtesy of The Meningitis Trust
(www.meningitis-trust.org).

Picture questions 3–6
Image is reproduced from *Sexually Transmitted Infections*, 2006; **82**:101–10; courtesy
of the BMJ Publishing Group.

Picture question 7
Photo is reproduced with permission from Island Sexual Health
(www.islandsexualhealth.org).

Picture question 8
Reproduced with permission from the University of Michigan Kellogg Eye Center
(www.kellog.umich.edu).

Picture question 9
Constable IJ. Age-related macular degeneration and its possible prevention. *MJA*,
2004; **181**: 471–72. © 2004, *The Medical Journal of Australia* – reproduced with
permission.

Picture question 10
© Brodie Paterson 2004. Reproduced from www.rcsed.ac.uk.

Picture question 12
Reproduced from www.ferengi.com.ar.

Picture question 13
© Crown copyright [2000–2005] Auckland District Health Board.

Picture question 14
Reproduced from www.arthritis.co.za.

Picture question 15
© Johns Hopkins University; reproduced from Withybush General Hospital PGMC
website (www.pdt-tr.wales.nhs.uk).

Picture question 16

Reproduced from www.baby-medical-questions-and-answers.com with permission.

Picture question 17

© 2007 Clinical Skills Education Centre, Queen's University Belfast.

Picture question 18

Reprinted with permission from eMedicine.com
(available at www.emedicine.com/emerg/TOPIC789.HTM).

Picture question 19

© www.curefootpain.co.uk – reproduced with permission.

Picture question 20

Reproduced from www.scienceblogs.com.

Picture question 21

Reproduced from www.flickr.com – photo by "mobiledoc".

Picture question 22

© 2007 Kevin T. Kavanagh; reproduced from www.entusa.com.

Picture question 23

Reproduced with permission from *The Merck Manual of Diagnosis and Therapy*,
Edition 18, edited by Mark H. Beers. © 2006 by Merck & Co. Inc., Whitehouse Station,
NJ. Available at www.merck.com/mmpe – accessed 14 January 2008.

Picture question 24

Image courtesy of visualdxhealth.com.

Picture question 25

Image courtesy of www.virtualmedicalcentre.com.

Picture question 26

© 2008 Memorial Sloan-Kettering Cancer Center. Reproduced from www.mskcc.org
with permission.

Picture question 27

© 2003–2008 Dermatology Online Atlas. Reproduced from www.dermis.net/doia/.

Picture question 28

© 2007 Dermaotology.co.uk. Reproduced from www.dermatology.co.uk with
permission.

Picture question 29

© 2006-07 University of Washington. Reproduced from HIV Web Study,
www.hivwebstudy.org.